John Robinson

The flora of Essex County, Massachusetts

John Robinson

The flora of Essex County, Massachusetts

ISBN/EAN: 9783337271336

Printed in Europe, USA, Canada, Australia, Japan

Cover: Foto ©berggeist007 / pixelio.de

More available books at **www.hansebooks.com**

THE FLORA

OF

Essex County, Massachusetts.

BY

JOHN ROBINSON,

SALEM:
ESSEX INSTITUTE.
1880.

TO

THE MEMORY OF

WILLIAM OAKES,

This little Flora of the County of Essex, where he
was born and where he loved to botanize,

Is Dedicated

BY THE
AUTHOR.

PREFACE.

The following enumeration of the plants of Essex County has been prepared, after a careful examination of the work of the earlier botanists and diligent search in almost every portion of the county for species not previously noticed.

With so few persons devoting themselves to the study of botany or the collection of specimens, particularly of the lower orders of plants, it would be impossible to present an absolutely complete list, and perhaps with even the greatest facilities no one has succeeded in so doing for any region.

Almost the only extended collection of dried specimens of county plants were those of the late Mr. Oakes, so that there really exists no very great foundation upon which to build other than the herbarium recently collected, and the writings of the more reliable among the earlier botanists, who for nearly a century have now and then appeared upon the scene.

Of the plants enumerated, almost all are represented in the herbarium of the Peabody Academy of Science at Salem, and where the species has not been collected and its occurrence is only known by the testimony of some writer, it is so stated in the list.

Several errors have been detected in early local lists and corrected, and such notes added to the paper as seem of interest locally or otherwise.

The writer can only in a general way here thank those who have assisted him in his work, and he trusts that the mention of their names in the following pages will be accepted as acknowledgments of his indebtedness.

PEABODY ACADEMY OF SCIENCE,
    SALEM, 1880.

Publications in which notices of interest relating to the Botanists or the Plants of the County may be found.

MEMOIRS OF THE AMERICAN ACADEMY OF ARTS AND SCIENCES, Vol. I. Boston, 1785. CUTLER'S LIST OF PLANTS.

BIGELOW'S FLORULA BOSTONIENSIS, 1st, 2nd and 3d editions.

CATALOGUE OF ANIMALS AND PLANTS OF MASSACHUSETTS in HITCHCOCK'S REPORT ON THE GEOLOGY, ETC., OF MASSACHUSETTS, 1833.

GRAY'S MANUAL, 1st, 4th and 5th editions.

GRAY'S FLORA OF NORTH AMERICA. Part I.

ESSEX INSTITUTE PROCEEDINGS, BULLETIN, AND HISTORICAL COLLECTIONS.

AMERICAN NATURALIST, BOTANICAL DEPARTMENT.

EMERSON'S TREES AND SHRUBS OF MASSACHUSETTS.

TUCKERMAN'S JOSSELYN'S NEW ENGLAND'S RARITIES DISCOVERED.

TRANSACTIONS OF AM. ANTIQ. SOCIETY, Vol. IV.

TRACY'S FLORA OF LYNN, etc.

PICKERING'S CHRONOLOGICAL HISTORY OF PLANTS.

WATSON'S REVISION OF THE LILIACEÆ IN PROC. AM. ACAD. ARTS AND SCIENCES. Vol. XIV.

WATSON'S BIBLIOGRAPHICAL INDEX TO N. A. BOTANY.

DECANDOLLE'S PRODROMUS (occasional reference to Oakes).

EATON'S FERNS OF NORTH AMERICA.

Harvey's Nereis Boreali-Americana.

Farlow's List of Marine Algæ of U. S. in Rept. Fish Com., 1875, and Proc. Am. Acad. of Arts and Sciences. Vol. X.

Flint's Grasses and Forage Plants.

Transactions of the Mass. Hort. Soc.

History of the Mass. Hort. Soc.

Oakes' Catalogue of Vermont Plants in Thompson's History of Vermont. Pages 173-208.

Halsted's Characeæ in Proc. Bost. Soc. Nat. Hist. Vol. XX.

List of the Plants of Georgetown and Vicinity, by Mrs. Horner, in Georgetown Advocate, 1876.

List of Plants near Danvers, by Dr. Geo. Osgood, in Salem Observer, 1853.

Hovey's Magazine of Horticulture (various articles by Oakes and Russell).

List of Plants of Pigeon Cove, by Calvin Pool, in "Pigeon Cove and Vicinity."

INTRODUCTORY.

INTRODUCTORY.

Essex County offers to the botanist a field attractive and interesting in many ways. The open country, deep woods, and numerous swamps contain the usual number of species found in such localities, while a large river, the Merrimac, furnishes a valley in which grow many plants not elsewhere found in the county. There are upwards of fifty ponds, from four to four hundred acres in extent, rich in water plants and subaquatics. Though there is no considerable hill or mountainous district, it is sufficiently far north to have several representatives of higher latitudes and even a few alpine and sub-alpine species in the flora.

Along the seashore is found an abundance of plants peculiar to the region of salt-water marshes and beaches, while in the ocean and inlets grow about one hundred and fifty species of algæ. These last named collecting grounds offer an opportunity to study, from fresh specimens, classes of plants from which the inland botanist is almost wholly debarred.

The land plants of the county belong decidedly to the northern flora although not so arctic in their character as the lichens and algæ. There is an almost total absence of many species common from Cape Cod southward and often found just south of Boston. In contrast to this the **Magnolia glauca** is still quite abundant at Gloucester, but not found again north of New Jersey. At

Cape Ann is the southern limit of the little **Sagina nodosa,** and there also is found **Potentilla tridentata,** familiar at the Isle of Shoals and on Mt. Washington. Essex County seems also to be the southern limit, for this region, of **Pinus resinosa** (Red Pine), **Abies nigra** (Black Spruce), **Vaccinium Vitis-Idæa, Viola rotundifolia,** etc., as it is the northern limit of **Cupressus thyoides** (White Cedar), **Quercus prinoides** (Chinquapin Oak), **Polygonum Caryi, Draba Caroliniana, Lygodium palmatum** (Climbing Fern), and others. At Boxford is what has proved thus far to be the only New England station for **Salix candida,** and another bog willow, **Salix myrtilloides,** is occasionally met with. At Andover a locality for **Calamagrostis Pickeringii** was discovered in the summer of 1879; this species has only been known before at the White Mountains. Among the sedges and grasses, plants too frequently neglected will be found, many not heretofore supposed to grow in the county, and a careful comparison of this list with our botanies will show that the range of many species has been extended. Although much careful work has been done there yet remains much to be accomplished; for, besides the few species that may be added to the list of flowering plants, there are many species of lichens and mosses not thus far collected, and the fungi and freshwater algæ are purposely omitted altogether. The phanerogams and vascular cryptogams are quite fully studied, and to the Characeæ and marine algæ but comparatively few additions may be expected.

The early settlement of the county renders this a particularly favorable region for the observation of introduced plants. From the earliest settlement to the present time, foreign species have continued to arrive, many of which, like the early colonists, came with the evident intention of remaining; for, as the genista, barberry, white-weed and buttercups show, they flourish here and increase to an extent which it would be difficult for them to exceed elsewhere. The study of these introduced plants might be called historical botany and should not be confounded with the study of the natural distribution and changes of plants. The early colonists came to establish a home: they did not come for gold, diamonds, or lead even, and in coming severed old home-ties and connections. That the fruit and other vegetable productions of the new land were among the first things to which attention was given, the records of early writers amply testify. We are apt to consider the men of two hundred and fifty years ago as a stern company; yet, besides the fruits and plants which might possess economic or medicinal value, this latter use being ever uppermost in the minds of botanical explorers of that day, they did not overlook the curious or the beautiful.

The earlier accounts tell of the gardens that were almost immediately established upon the settlement of the country, and invoices of the articles to be sent to the colonists from the managers in Europe contain such things as the seeds of grains, stone fruits, quince, apple, pear, woadwax, barberry, etc. Besides these, living plants must have been sent out from Europe, as is shown by the record of "Our Ancient Pear Trees" (Robert Manning in Proc. Am. Pom. Soc., 1875).

Some of these plants purposely introduced have failed to prove of use, or their time of usefulness has gone by, and they have been suffered to run wild, and at the same time a hundred others have like "stowaways" come uninvited. They have been introduced among the seeds of useful plants, in packing material, and as garden flowers. Many of the introduced species still remain restricted to certain localities, and others, although more widely disseminated, are in such situations as to make their origin self-evident, while others are so distributed as to appear to all intents and purposes as natives. Again, by the clearing of the forests, the general cultivation and changes in the condition of the soil, many native plants best able to endure the changes, or those to which the changes have proved beneficial, have been given positions of undue prominence in the flora; while other species, which at the time of the settlement of the country were much more abundant, have now become less numerous, or have entirely disappeared. It is a matter of considerable difficulty to picture to ourselves the country as it appeared two hundred and fifty years ago. It is probable that extensive forests reached to the ocean shore and, excepting the river marshes and clearings made by the fires of the aborigines, occupied the whole territory. The Indians cultivated corn, pumpkins, beans, tobacco and a few other plants. It is possible that some species of foreign plants had been introduced previous to the settlement by the whites through trade or by adventures, but this is uncertain. The study of the introduced plants is aided by the work of Mr. John Josselyn (New England Rarities Discovered), a reprint of which, with valuable notes by Professor Edward Tuckerman, is now available. Josselyn visited New England several times; when on the longest sojourn, 1663–1671, he landed at Boston and soon went to Black Point, Scarborough, Maine, where most of his observations were made. Josselyn was an excellent observer and although his writings are filled with the usual strange stories current in old works upon new and unexplored countries, they contain the first accounts of any consequence regarding the New England flora. This author did not, perhaps, make many observations in Essex County, yet his work contains but few species that do not grow here and its chief value consists in its arrangement and separation of the plants indigenous from the introduced weeds, thus giving what then appeared to be the plants which came with man or, as he called them, "Such plants as have sprung up since the English planted and kept cattle in New England." This, with the occasional observations by other writers, gives us a fair idea of what plants had established themselves here rather more than two hundred years ago. According to Professor Tuckerman, the next date by which the student may fix the introduction of foreign species is 1783, when the list of plants observed by Rev. Manasseh Cutler, of Ipswich, was published (Mem. Am. Acad. Vol. I). Since that date observations are more frequent and the more recently introduced species can be traced quite accurately. It is also quite probable that plants which at one time were quite common weeds have disappeared altogether. Dr. Cutler mentions the **Amarantus** known by the common name of "Prince's Feather" or "Love-lies-bleeding," as "amongst rubbish," but to the present writer's knowledge it is never met with excepting in old-fashioned gardens. The **Hyoscyamus niger** and **Artemisia Absinthium** (Wormwood), spoken of by Dr. Cutler and other earlier writers as common in waste places, are now very rare or unknown. The last mention of **Nicotina rustica** is by Dr. Osgood in 1853, but it is doubtful if he observed it as late as that; his observations were very probably made in previous years, and no one has since noticed it. The introduction of new manufactures is likely to bring with it plants which may be persistent enough in the region where they are introduced but unknown elsewhere. Such is the case at "Tapleyville," Danvers, where, in the vicinity of a carpet factory established forty years ago, are to be found several species of foreign plants unknown in any other town of the county, and perhaps not elsewhere established.

Two or three plants observed along the shore of the Merrimac river suggest a close connection with the mills at Lowell and Lawrence, one of them being a southern sedge. Many plants are emigrating eastward from our western states, travelling as it were by rail. The **Rudbeckia hirta,** now quite common in fields hereabouts, according to Dr. Pickering, did not reach Philadelphia until 1826, and this vicinity until perhaps 1855.

The latest arrival noticed (1878) is that of **Eleusine Indica,** a weedy, oriental grass which is common at New York city and Philadelphia. It has made its appearance along the railroad tracks at the Pennsylvania Pier, Salem, having travelled thence by the P. and R. R. R. Co's steamers, which regularly bring coal from Philadelphia. This last comes under the head of "ballast plants," a very full account of which may be found in the Torrey Bulletin for November, 1879.

SKETCH OF SOME
OF
THE EARLY BOTANISTS.

SKETCH OF SOME
OF
THE EARLY BOTANISTS.*

THE study of botany in Essex County, we may in fact say New England, dates from the time of Dr. Manasseh Cutler at the close of the last century. Previously the plants had only been noticed by writers upon more general subjects of natural history, or casually mentioned in letters written from this country to England. But from Cutler's time there has been a steady succession of botanists, chiefly amateurs, who have kept alive an interest in the subject, even at times making it the prominent topic considered at the literary and scientific societies and clubs of the region. It will only be attempted here to give a brief sketch of the older botanists who have contributed most to the knowledge of the subject in the county.

Francis Higginson, in a letter written from Salem in 1629-30 (Mass. Hist. Coll., I, 121), speaks of the plants which he had noticed growing in the vicinity, and mentions several species which probably now exist in the same localities as observed by him at that early date; one, the *Rubus odoratus* (Flowering Raspberry or Mulberry) still flourishes in the "Great Pastures," and the *Osmorrhiza longistylis* (Chervil or Sweet Cicely) has been noticed until very recently at "Paradise," near Salem.

William Wood, in the New England Prospect, speaks

---
* The writer is indebted to Dr. Henry Wheatland for his assistance in obtaining notices of the early botanists of the county, chiefly from the Proceedings and Historical Collections of the Essex Institute, from which a large portion of this sketch is made.

extendedly of the early gardens and the numerous useful plants native to the country, mentioning what he saw at Ipswich, Salem, Marblehead, etc.; Parkinson and Jerard enumerate New England plants; John Josselyn, previously referred to, gives an account of the native and introduced species; and other early writers, including John Winthrop, speak of the excellent quality of the native fruits and the beauty of the flowers, particularly dwelling on the superiority and abundance of the wild strawberries.

None of these can, however, be spoken of or claimed as Essex County botanists, and it is not until after the close of the American Revolution that we find any serious or scientific study of the plants of the county.

Dr. Manasseh Cutler was born at Killingly, Connecticut, May 3, 1742, graduated at Yale College in 1765, afterward studied law, and was admitted to the bar in 1767. He soon studied for the ministry and was settled at the Hamlet Parish in Ipswich, which was set apart from that town and named Hamilton for Alexander Hamilton whom Dr. Cutler greatly admired. He served as a chaplain during the war of the revolution and on his return studied medicine which he afterwards practised among his parishioners. The efforts of Dr. Cutler secured the passage, in 1787, of the famous ordinance by which freedom was declared in the northwestern territories and he soon after organized the first band of pioneers that emigrated from the east to Ohio. The next year he followed them driving himself the entire distance in a sulky, being accompanied by a few friends. Upon his return from the west, or in 1800, he was chosen to represent old Essex in Congress where he served two terms. While in Philadelphia in 1787, he visited at the house of Benjamin Franklin, and afterward wrote an account of the great statesman which was considered as one of the best, being copied by Sparks in his life of Franklin. Dr. Cutler prepared, in 1783, "An account of some of the vegetable productions, naturally growing in this part of America, botanically arranged," which was published in the first volume of the Memoirs of the American Academy of Arts and Sciences in 1785. He here described some three hundred and fifty species of flowering plants suggesting several points which have been followed by later botanists. It was Dr. Cutler's intention to extend this work, and there are in existence several manuscript volumes which he prepared toward this end. These valuable manuscripts are in the possession of Prof. Edward Tuckerman, who intends that their final destination shall be the library of Harvard; and it is to be hoped that they may at some future day be printed, with such notes as would be required to make them of use to the present generation of botanical students. Dr. Cutler's death occurred in 1823, after more than fifty years' service in one parish. He has been called the father of American botany, a term certainly appropriate for the times and for this region, where his mantle fell on the shoulders of Osgood, Nichols, Oakes, and Pickering.

Dr. George Osgood, son of George and Elizabeth (Otis) Osgood, was born at Fair Haven, March 25, 1784. He studied medicine with his father and settled in Danvers in 1804, where he had for many years an extensive practice. Dr. Osgood acquired, by his association with Cutler, Nichols and Oakes, a taste for and knowledge of botany which lasted him through life. He contributed to Dr. Bigelow much valuable information while the latter was preparing his "Florula Bostoniensis," and in 1853 published in the Salem Observer a local list of flowering plants. He died May 26, 1863.

Dr. Andrew Nichols was born in the rural part of Danvers, Nov. 22, 1785. He was the son of Andrew and Eunice (Nichols) Nichols, and studied medicine under Dr. Waterhouse, settling in that part of Danvers, now Peabody, in 1808, where he practised successfully, remaining there until his death, March 31, 1853.

He was particularly interested in the local natural history of this region, and in 1816 delivered a series of lectures on botany, the first of such in this part of the country. Dr. Nichols was one of the founders of the Essex County Natural History Society and its president, retaining unabated till death his interest in his favorite study.

William Oakes must be acknowledged as the most eminent botanist of Essex County birth. He was the son of Caleb Oakes and was born at Danvers, July 1, 1799. He was educated at Harvard receiving the degree of A. B. in 1820. He early developed a taste for natural history relinquishing the practice of law, his chosen profession, to study this branch of science.

Mr. Oakes' work was chiefly in New England, collecting extensively in Essex County, Mass., Vermont, the White Mountain region, and southeastern and western Massachusetts. He prepared the list of plants of Vermont for Thompson's history of that state; and his work at the White Mountains was so thorough that recent collectors, with all the advantages of improved roads and easy access to every portion of that region, have failed to add but few to the number of species which he discovered there. It was his intention to have published a flora of New England, but was deterred by the appearance of Beck's Botany. He afterwards became deeply interested in a work, with illustrations by Sprague, upon White Mountain scenery, which was published in 1848; but not until after his death which occurred July 31, 1848, the preface of the work having been written July 26, only five days previous.

Mr. Oakes was impulsive and generous; thoroughly in earnest in his favorite study, he seriously impaired his fortune to carry out his schemes more perfectly. Like many other men of note, he was hardly appreciated while living, but no monument which could have been erected would have made his memory more cherished or his worth more appreciated by the present generation of botanists than that which he left behind,— an extensive collection of most beautifully prepared botanical specimens, with an identification absolutely correct, besides many valuable notes and observations. Prof. Tuckerman dedicated to him a pretty little plant common in the region of Plymouth, but it afterwards had to be transferred to another genus; and now for the first time in any flora, it becomes a pleasant duty to give by its name, "Oakesia," the little bellwort, a common Essex County plant, which Prof. Watson of Cambridge has found necessary to separate from the genus to which it has heretofore been referred in his revision of the family Liliaceæ, and has feelingly dedicated to the memory of William Oakes.

Dr. Charles Pickering, son of Timothy and Lurena (Cole) Pickering and grandson of Col. Timothy Pickering of revolutionary fame, was born at Starucca Creek on the Susquehanna, Pennsylvania, in 1805. He was educated at Harvard in the class of 1823, graduating at the medical school in 1826. In 1838 he was appointed naturalist to the U. S. (Wilkes) Exploring Expedition; and to perfect his knowledge of animals and plants in foreign parts, he made very extensive journeys after his return from that expedition. He was the author of several works of great value which in their preparation required much untiring research; among them are "Geographical distribution of

Animals and Plants" and "Chronological History of Plants," the latter work occupying the last sixteen years of his life in its preparation.

During his college life Dr. Pickering spent much of his time at Wenham, at the homestead of his grandfather, Col. Pickering, and here he was in the habit of botanizing in company with William Oakes, a favorite locality being the "Great Swamp." It is but right that Essex County should claim a share of the honor of his name, for it was here that his attention was drawn to botany, and in the Chronological History of Plants, page 1063, we find the following entry "1824 * * In this year, after an excursion in 1823, with William Oakes diverting my attention from entomology, my first botanical discovery." Dr. Pickering died at Boston, March 17, 1878. The writer will always remember with pleasure and gratitude the many hours spent with Dr. Pickering during 1876 and '77, while he patiently sought out, among his early manuscript notes and his letters from William Oakes, the species and stations noticed while botanizing in Essex County more than fifty years before.

Rev. John Lewis Russell, son of John and Eunice (Hunt) Russell, was born at Salem, Dec. 2, 1808. He was at Harvard in the class of 1828, and graduated at the divinity school in 1831. After occupying pulpits in Chelmsford, Hingham, Brattleboro, Kennebunk and various other places, he returned in 1853 to Salem, where he resided, preaching occasionally, until his death June 7, 1873.

Mr. Russell was particularly devoted to cryptogamic botany, publishing accounts of his investigations from time to time as he proceeded, besides many popular articles on various families of plants. He lectured frequently on botany and was for many years vice-president of the Essex Institute.

Mr. Russell contributed much to the general knowledge of botany in Essex County, but his most extensive collections were made in other places.

The only attempt at an enumeration of county plants, as such, is that of Mr. Cyrus M. Tracy, of Lynn. It was intended to give a list of the flowering plants found in that region and contained 546 species. Besides possessing a very happy gift as a botanical lecturer, Mr. Tracy has contributed several valuable articles upon local botany to the publications of the Essex Institute and elsewhere.

Mr. Geo. D. Phippen, of Salem, whose notes on the native plants have materially aided the writer, has often presented the subject of botany at meetings of the Institute, and has written several articles of interest upon the subjects which have been published in various places. Mrs. C. N. S. Horner, of Georgetown, a most excellent botanical collector, published a list of the plants of that region in the Georgetown Advocate in 1876. Mr. Calvin Pool, of Rockport, prepared a somewhat smaller list of plants of Cape Ann, which was published in "Pigeon Cove and vicinity" in 1873. Mr. S. B. Buttrick, whose years do not diminish his interest in botany, and who is ever on the alert to find some rare flower, has contributed several lists of plants to the earlier numbers of the Proceedings and Bulletin of the Essex Institute, as also have Dr. G. A. Perkins, of Salem, chairman of the botanical section of the Peabody Academy of Science, Mr. Geo. F. H. Markoe, now of Boston, Rev. Mr. Chute, formerly of Lynnfield, and many others. Dr. Henry Wheatland, although not claiming to be a botanist, has often aided those who did, by his assistance in revising the articles for the publications of the Institute while acting as the editor. Mr. S. P. Fowler, of Danvers, one of the older botanists and a companion of Oakes and Osgood, in many rambles, has made frequent observations regarding the trees and shrubs, and has cultivated extensively many of our native plants. Of those who have contributed to swell the list of known county species of plants and who have not published any writings on the subject, it will be impossible to speak separately. They must be content to feel that they have aided the cause of botanical knowledge as they certainly have, and are deserving their share of credit for so doing. There are many who have collected and prepared specimens which will always serve as pleasant reminders of their work. Among such are Mr. B. D. Greene, who added several plants to the flora from near Tewksbury; Mr. Wm. P. Richardson and Mr. S. Bass, who botanized near Salem; and more recently Mrs. Alex. Bray, Mrs. Charles Grover, Mrs. J. Babson and Mrs. Davis, who have added many species to the list of Cape Ann Algæ; Mr. Frank Lufkin of Pigeon Cove, who has noticed many plants new to that region; Mr. John H. Sears, of Danvers, whose specialty, the forest trees, has been the means of bringing together at the Museum of the Peabody Academy of Science one of the best local collections of native woods in any museum; Mr. W. P. Conant, who has added many species of Cyperaceæ and Gramineæ and a rare Botrychium to the flora; besides many others whose names will appear in the list associated with the plant which they have been fortunate to discover.

To those whose assistance either by their writings, by specimens contributed, or who have rendered any assistance by information or other attention, the writer desires to express his warmest thanks. It would be impossible for him to specify those who have aided him or their manner of so doing, and he can only thank all collectively which he does most sincerely.

FLORA.

FLORA.

EXOGENS.

RANUNCULACEÆ.

(CROWFOOT FAMILY.)

Clematis Virginiana, *L.* (VIRGIN'S BOWER.)
Common. Climbing over bushes and low trees.

Anemone cylindrica, *Gray.* (LONG-FRUITED ANEMONE.)
Legge's Hill (Rev. J. L. Russell); Andover (Mrs. Downs); etc. Not very common.

Anemone Virginiana, *L.*
Chebacco, Andover, Haverhill, etc. Frequent.

Anemone nemorosa, *L.* (WOOD ANEMONE.)
Very common among bushes in moist places.

Anemone Hepatica, *L.,* Hepatica triloba, *Chaix.* (Gray's Manual). (HEPATICA.)
Found January 16, 1871, opening its flowers after a few warm days. An unusual occurrence. Dry hard woods. Common.

Thalictrum anemonoides, *Michx.* (RUE ANEMONE.)
Dry woods. Not uncommon.

Thalictrum dioicum, *L.* (EARLY MEADOW RUE.)
Common in rocky places.

Thalictrum purpurascens, *L.*
This species is found near Boston (C. E. Faxon), and very probably grows in the county, but has not yet been collected.

Thalictrum Cornuti, *L.* (TALL MEADOW RUE.)
Common.

Ranunculus aquatilis, *L.,* var. **trichophyllus,** *Chaix.* (COMMON WHITE WATER-CROWFOOT.)
Wenham, Chebacco and other ponds, but rarely in flower.

Ranunculus multifidus, *Pursh.* (YELLOW WATER-CROWFOOT.)
Middleton, Essex, etc. Not rare.

Ranunculus ambigens, *Watson,* Ranunculus alismæfolius, *Geyer* (Gray's Manual).
Collected by W. P. Conant not far beyond the county limits in New Hampshire. It is probable that this species will be found nearer.

Ranunculus Flammula, *L.*, var. **reptans.** (CREEPING SPEAR-
WORT.)
Common along the shores of ponds and rivers.
Ranunculus Cymbalaria, *Pursh.* (SEASIDE CROWFOOT.)
Nahant, Swampscott, Danvers, Gloucester, and other places near
the coast.
Ranunculus abortivus, *L.* (SMALL FLOWERED CROWFOOT.)
Common in damp shady places.
Ranunculus recurvatus, *Poir.* (HOOKED CROWFOOT.)
Near Salem, 1857 (S. B. Buttrick, Proc. E. I., Vol. II, p. 235);
Essex County (Oakes); Oakes' ledge, Danvers, 1879 (J. R.). Rare.
Ranunculus sceleratus, *L.* (CURSED CROWFOOT.)
Nahant; Calf spring (Tracy); Middleton (Osgood, memo. Rev.
J. L. Russell); Essex County (Oakes); Plum Island; Salem Great
Pastures. Not very common.
Ranunculus Pennsylvanicus, *L.* (BRISTLY CROWFOOT.)
Lawrence, above the dam, north side, Aug., 1879 (J. R.).
Ranunculus fascicularis, *Muhl.* (EARLY CROWFOOT.)
Andover (Mrs. S. M. Downs); near Danvers (Dr. Osgood's list).
Ranunculus repens, *L.* (CREEPING CROWFOOT.)
Danvers, near the trotting park (J. R.); Andover (Mrs. Downs).
Wet places.
Ranunculus bulbosus, *L.* (BUTTERCUPS.)
Very common, earlier than the next species. (Nat. from Eu.)
Ranunculus acris, *L.* (TALL BUTTERCUPS.)
About as common as last. (Nat. from Eu.)

Caltha palustris, *L.* (MARSH MARIGOLD.)
Often misnamed "Cowslips," which is the common name given to a
Primrose (Primula officinalis) in England. Frequent in meadows.

Coptis trifolia, *Salisb.* (GOLDTHREAD.)
Abundant in moist woods.

Aquilegia Canadensis, *L.* (WILD COLUMBINE.)
Very common on rocky hillsides.
Var. **Phippenii.**
Flowers salmon colored, leaves lighter green; transplanted to the
garden it seeded freely and invariably produced its like. Discovered
by Mr. G. D. Phippen in a ravine in Salem pastures about 1844.
Found again in the same locality by the present writer, 1875, and
by Mr. David Waters in 1880. A white variety was detected by
Mr. Abraham Bosson among red Columbines, but did not prove
hardy on being transplanted. (See "Notice of three varieties of
native Columbines," Proc. E. I., Vol. I, p. 268).

Aquilegia vulgaris, *L.*
The garden Columbine occasionally is found escaped.

Actæa spicata, *L.*, var. **rubra,** *Michx.* (RED BANEBERRY.)
In many parts of the county, but rather scarce.

Actæa alba, *Bigelow.* (WHITE BANEBERRY.)
In similar situations, but more common than the last. Moist rocky places or by shady walls.

Cimicifuga racemosa, *Ell.* (BLAKE SNAKEROOT.)
Collected along the embankment of the railroad near West Gloucester some years since (memo. Mr. G. D. Phippen). It is probable that the plants were introduced although they were at the time quite abundant.

MAGNOLIACEÆ.

(MAGNOLIA FAMILY.)

Magnolia glauca, *L.* (SMALL MAGNOLIA.)
Gloucester and swamps towards Essex. First brought to notice by Rev. Manasseh Cutler during the last century.

Liriodendron Tulipifera, *L.* (TULIP TREE.)
Introduced from the west and south, as are some species of Magnolia, as an ornamental tree.

BERBERIDACEÆ.

(BARBERRY FAMILY.)

Berberis vulgaris, *L.* (COMMON BARBERRY.)
Very abundant in rocky places and along walls. (Nat. from Eu.)

Caulophyllum thalictroides, *Michx.* (BLUE COHOSH.)
Georgetown (Mrs. C. N. S. Horner). Rare.

Podophyllum peltatum, *L.* (MAY APPLE MANDRAKE.)
Extensively introduced in old gardens, and said by some to be a native of the county, but this is very doubtful. Of this plant Gray says that the fruit is edible while the leaves and roots are drastic and poisonous.

NYMPHÆACEÆ.

(WATER-LILY FAMILY.)

Brasenia peltata, *Pursh.* (WATER-SHIELD.)
Quite common in most ponds and in slow places in the rivers. It would seem that the gelatinous substance which abounds on all submerged portions of the plant might be made of use.

Nymphæa odorata, *Ait.* (WATER-LILY.)
Common in ponds and slow streams. Very pink flowers are collected in shallow ponds at Gloucester and Danvers, but not to be compared to the pink lilies from Barnstable, Mass.

Var. **minor,** *Sims.*
Apparently nothing more than a small form of the common pond-lily with smaller leaves and flowers, the petals and sepals being more obtuse. Common in Chebacco and some other ponds.

Nuphar advena, *Ait.* (YELLOW POND-LILY; COW-LILY.)
Very abundant. The small nuphar ought to be found in some of our ponds.

SARRACENIACEÆ.

(PITCHER-PLANT FAMILY.)

Sarracenia purpurea, *L.* (PITCHER-PLANT; SIDE-SADDLE FLOWER.)
Quite common in bogs. An interesting variety with yellowish green flowers; has been found in Beverly for a number of years by Wm. D. Silsbee. The writer found an abnormal condition of this plant at North Reading, June, 1872, bearing two flowers consisting of many rows of sepals, one circle within another.

PAPAVERACEÆ.

(POPPY FAMILY.)

Chelidonium majus, *L.* (CELANDINE.)
A common weed near dwellings. "Introduced from Europe prior to 1669 (see Josselyn Rar.)" (Dr. Charles Pickering, Chron. Hist. Pl., p. 242).

Sanguinaria Canadensis, *L.* (BLOOD-ROOT.)
Frequent in rocky and shady places.

Argemone Mexicana, *L.* (PRICKLY POPPY.)
On the road to Flax Pond, Lynn, 1879 (Herbert A. Young). Introduced from tropical America. Not permanently established.

FUMARIACEÆ.

(FUMITORY FAMILY.)

Adlumia cirrhosa, *Raf.* (CLIMBING FUMITORY.)
Introduced into gardens from the west. It freely perpetuates itself by seeds and is now and then found escaped.

Dicentra cucullaria, *DC.* (DUTCHMAN'S BREECHES.)
Gloucester, 1863 and 1877 (Mrs. Babson); "Andover" (memo. Mrs. Downs). Rare.

Corydalis glauca, *Pursh.* (CORYDALIS.)
Rocky hills. Not very common.

Fumaria officinalis, *L.* (COMMON FUMITORY.)
Ipswich, Danvers (Oakes, 1819); Wenham (Miss Davis); Salem (S. B. Buttrick); Boxford (Miss Perley). A rather scarce escaped plant. (Int. from Eu.)

CRUCIFERÆ.

(MUSTARD FAMILY.)

Nasturtium officinale, *R. Br.* (EUROPEAN, OR TRUE WATER CRESS.)
"Lynn" (Tracy's list); "Andover" (memo. Rev. H. P. Nichols). It is very doubtful if this species grows in the county, as the cress usually found is Cardamine hirsuta.

Nasturtium palustre, *DC.* (MARSH CRESS.)
In wet places. Quite common and variable.

Nasturtium Armoracia, *Fries.* (HORSERADISH.)
Often escaping from cultivation. (Introduced from Europe.)

Cardamine rhomboidea, *DC.* (SPRING CRESS.)
Not rare in wet places.

Cardamine hirsuta, *L.* (SMALL BITTER CRESS.)
Brooks and other wet places. Common. Sold in the market as Water Cress and often supposed to be N. officinalis.

Var. sylvatica.
"Lynn" (Tracy); "Essex County" (memo. Dr. Charles Pickering.)

Arabis lævigata, *DC.*
Beverly (John C. Phillips).

Arabis Canadensis, *L.* (SICKLE-POD.)
Topsfield (Oakes); Danvers (J. H. Sears); and a few other localities. Rather scarce.

Arabis perfoliata, *Lam.* (TOWER MUSTARD.)
"Paradise," Salem, 1819 (S. Bass); Essex County (Oakes, memo. Hovey's Mag., Vol. VII); near the Ocean House, Revere Beach (C. E. Faxon).

Barbarea vulgaris, *R. Br.* (YELLOW ROCKET.)
Quite common in damp places.

Sisymbrium officinale, *Scop.* (HEDGE MUSTARD.)
A common weed. (Nat. from Eu.)

Sisymbrium arvense.
Found by Rev. J. L. Russell, with other weeds, persisting for many years, in Salem. (Adv. from Europe.)

Brassica alba. (WHITE MUSTARD.)
Found on the South Boston "dump" (C. E. Faxon); also reported as growing in Andover (Mrs. Downs). (Introduced from Europe.)

Brassica nigra. (BLACK MUSTARD.)
In various parts of the county. (Adv. from Europe.)

Brassica campestris, *L.* (CABBAGE, ETC.)
Var. **Colsa.** (RAPE.)
Found in a street in Salem.
Var. ———
A tall smooth form found in cultivated fields in Danvers (J. H. Sears). (Introduced.)

Draba Caroliniana, *Walt.*
Salem 1824 (Dr. Chas. Pickering). The original locality still exists, the plants being more or less abundant each year, usually appearing in April.

Draba verna, *L.* (WHITLOW GRASS.)
Danvers 1826 (Dr. Andrew Nichols). This species is only found at the old locality, and some seasons but two or three plants are to be seen, yet it has retained its place in our flora, now and then appearing in comparative abundance; if a locality, the extent of which is hardly more than ten feet in diameter, can be considered as entitled to that term.

Alyssum maritimum, *L.* (SWEET ALYSSUM.)
Common in gardens often escaping in yards and streets. (Nat. from Eu.)

Camelina sativa, *Crantz.* (FALSE FLAX.)
On Derby Wharf, Salem, 1877. Introduced by the colonists continuing a weed, observed (probably 1824) "at Salem" (Dr. Chas. Pickering, Chron. Hist. Pl. p. 353); Boxford (Miss Perley).

Capsella Bursa-pastoris, *Mœnch.* (SHEPHERD'S PURSE.)
A most common weed. In flower from May to November. (Nat. from Eu.)

Thalaspi arvense, *L.* (FIELD PENNYCRESS.)
North Salem, 1858 (Geo. P. Bradford, Proc. E. I., Vol. II, p. 237); "Yard of Plummer Hall, 1859" (memo. Rev. J. L. Russell). (Nat. from Eu.)

Lepidium Virginicum, *L.* (WILD PEPPERGRASS.)
Common in fields and along roadsides.

Lepidium campestre, *L.*
Eastern R. R. road-bed, North Beverly. (Adv. from Eu.)
Lepidium ruderale, *L.*
Common about Boston extending undoubtedly into Essex Co. (memo. C. E. Faxon.) Noticed by Dr. Pickering about 1824. (Nat. from Eu.)

Cakile Americana, *Nutt.* (SEA-ROCKET.)
Common along the shore.

Raphanus Raphanistrum, *L.* (JOINTED CHARLOCK.)
A common weed in fields. (Adv. from Eu.)
Raphanus sativus, *L.* (GARDEN RADISH.)
Occasionally by the roadside. Hamilton, 1875, etc. (Int. for cult.)

VIOLACEÆ.

(VIOLET FAMILY.)

Viola rotundifolia, *Michx.* (ROUND-LEAVED VIOLET.)
Gloucester (Mrs. J. Babson); Rockport (C. W. Pool). Rare.
Viola lanceolata, *L.* (LANCE-LEAVED VIOLET.)
Common in wet places.
Viola primulæfolia, *L.* (PRIMROSE-LEAVED VIOLET.)
Frequent in moist land.
Viola blanda, *Willd.* (SWEET WHITE VIOLET.)
Common. Gray considers these three species as connected directly with each other, which certainly seems to be the case with our county specimens.
Viola odorata, *L.* (ENGLISH VIOLET.)
Salem (old gardens), now and then escaping by the roadside.
Viola cucullata, *Ait.* (SWAMP VIOLET.)
Common. Varies greatly both as to the size and color of the flowers and shape of the leaves. "Under cultivation, leaves greatly enlarged with palmate and cordate on the same plant." (Memo. G. D. Phippen). A variety with variegated blue and white flowers has been three times transplanted, still retaining its characteristics.
Viola sagittata, *Ait.* (EARLY BLUE VIOLET.)
Very common. Sometimes in damp places resembling the last.
Viola pedata, *L.* (PEDATE VIOLET.)
Not rare. Beverly, Hamilton, Lynn, etc., but most abundant along the Merrimac valley.
Viola canina, *L.*, var. **Sylvestris,** *Regel.* (DOG VIOLET.)
Quite common in moist fields.

Viola pubescens, *Ait.* (YELLOW VIOLET.)
Rockport (C. W. Pool); Danvers; Andover; Georgetown (Mrs. Horner). Rather scarce. Varies considerably, some specimens being very downy while others are quite smooth.

Viola tricolor, *L.* (HEART'S-EASE. SMALL PANSY.)
Frequently escapes from old gardens. "Roadsides out of town from Ipswich" (memo. G. D. Phippen). (Int. from Eu.)

CISTACEÆ.

(ROCK-ROSE FAMILY.)

Helianthemum Canadense, *Michx.* (FROST-WEED.)
Quite common in pasture lands.

Hudsonia tomentosa, *Nutt.*
Nahant, Ipswich, Salisbury, Plum Island, Coffin's Beach, West Gloucester, etc. Growing in sand, it often forms little hills in its efforts to overtop the sand which blows over it.

Lechea major, *Michx.*
Quite common in pastures.

Lechea thymifolia, *Pursh.*
Abundant at Ipswich with Hudsonia, but rather scarce in the interior.

Lechea tenuifolia, *Michx.*
Dry hills. Common.

Lechea intermedia (Prov.), *Leggett.* Perhaps a form of L. minor.
Common in dry pastures.

Lechea minor, *Lam.*
Dry pastures. Frequent.

DROSERACEÆ.

(SUNDEW FAMILY.)

Drosera rotundifolia, *L.*
Quite common in meadows, and wet paths in the woods.

Drosera intermedia, *Drev. & Hayne,* var. **Americana,** *DC.*
Drosera longifolia, *L.* (Gray's Manual.)
Meadows. More common than the last. These interesting plants are treated very elaborately by Mr. Darwin in his volume on insectivorous plants.

HYPERICACEÆ.

(St. John's-wort Family.)

Hypericum ellipticum, *Hook.*
Wenham swamp, 1824 (Dr. Charles Pickering); Danvers, 1877; Boxford (Miss M. E. Perley.) Rare.

Hypericum perforatum, *L.* (Common St. John's-wort.)
Common. A pretty weed from Europe.

Hypericum corymbosum, *Muhl.*
Georgetown (Mrs. Horner); Haverhill, etc. Scarce.

Hypericum mutilum, *L.*
Very common in wet places.

Hypericum Canadense, *L.*
Shores of ponds and meadows. Very common.

Hypericum Sarothra, *Michx.* (Pine-weed; Orange grass.)
Railroad tracks, etc. Common.

Elodes Virginica, *Nutt.* (Marsh St. John's-wort.)
Quite common in swampy places.

ELATINACEÆ.

(Water-wort Family.)

Elatine Americana, *Arnott.*
"Beaver pond, 1869" (memo. Rev. J. H. Russell); Crane pond, West Newbury, 1879 (J. R.); Flax pond, Lynn (E. Faxon).

CARYOPHYLLACEÆ.

(Pink Family.)

Dianthus barbatus (Sweet William.)
Danvers, escaped for a number of years (J. H. Sears). (Adv. from Eu.)

Dianthus Armeria, *L.* (Deptford Pink.)
Quite common in the vicinity of old towns, particularly Salem. (Adv. from Eu.)

Dianthus deltoides, *L.*
Appeared in West Newbury as an escaped plant in 1878 (W. P. Conant). Prof. Watson, who kindly looked up the species, states it has also been sent from Plymouth, N. H. (Adv. from Eu.)

Saponaria officinalis, *L.* (Common Soapwort; Bouncing Bet.)
Common. Both the single and double flowering varieties are frequent near old gardens, and along roadsides. (Adv. from Eu.)

Silene inflata, *Smith.* (BLADDER CAMPION.)
Very common in Lynn and Swampscott on the road-bed of the Eastern R. R. and in other towns more or less abundant. (Nat. from Eu.)

Silene Pennsylvanica, *Michx.* (WILD PINK.)
Reported at Danvers, Andover, and occasionally in other parts of the county.

Silene Armeria, *L.* (SWEET WILLIAM CATCHFLY.)
Quite common near Salem, Lynn and also in some other places. (Adv. from Eu.)

Silene antirrhina, *L.* (SLEEPY CATCHFLY.)
Common in dry places on poor soil.

Silene noctiflora, *L.* (NIGHT-FLOWERING CATCHFLY.)
Gloucester, Rockport, Salem. Rather scarce.

Lychnis Githago, *Lam.* (CORN COCKLE.)
Danvers, Salem, Lynn. Scarce. (Adv. from Eu.)

Lychnis vespertina, *Sibth.* (EVENING LYCHNIS.)
Andover (memo. Rev. H. P. Nichols). Escaped from cultivation.

Arenaria serpyllifolia, *L.* (THYME-LEAVED SANDWORT.)
Quite common on poor lands. (Nat. from Eu.)

Arenaria lateriflora, *L.*
Frequent in damp places.

Arenaria peploides, *L.*
Rockport (Frank Lufkin); Salisbury beach (Mrs. Downs); "Kings beach and others in that region" of Lynn (Tracy); Ipswich (Oakes).

Stellaria media, *Smith.* (CHICKWEED.)
Everywhere, often in flower under the snow in winter. (Nat. from Eu.)

Stellaria longifolia, *Muhl.* (LONG-LEAVED STICHWORT.)
"Near the outlet of Cedar pond, Lynnfield" (Dr. Chas. Pickering); Pickman Farm, Salem (J. R.).

Stellaria borealis, *Bigelow.*
Lynnfield (A. P. Chute, Proc. E. I., Vol. II, p. 47). This is not represented in the county collection at the P. A. S. Doubtful.

Cerastium viscosum, *L.* (MOUSE-EAR CHICKWEED.)
A common weed. (Nat. from Eu.)

Cerastium arvense, *L.* (FIELD CHICKWEED.)
Nahant; Gloucester (Mrs. J. Babson); Rockport (C. W. Pool). Not rare near the coast.

Cerastium vulgatum, *L.* (MOUSE-EAR CHICKWEED.)
Amesbury and Salisbury, June 25, 1863 (Markoe and Shepard, Proc. E. I., Vol. III, p. 272); "Andover" (memo. Mrs. Downs). Not

represented in the P. A. S. collection, and very doubtfully a county species.

Cerastium nutans, *Raf.*, which finds a place in some local printed lists, certainly must be an error.

Sagina procumbens, *L.* (PEARLWORT.)
Common in damp places. A weed in among the bricks of door yards in Salem.

Sagina nodosa, *Fenzl.*
Pigeon Cove, 1875 (J. R.). Rare.

Lepigonum rubrum, *Fries,* var. **campestris.** Spergularia rubra, *Presl,* var. campestris. (Gray's Manual.) (SAND-SPURRY.)
Common along roadsides near the shore.

Lepigonum medium, *Fries.* Spergularia media, *Presl.* (Gray's Manual.)
All salt marshes. It is quite difficult to separate these plants in a satisfactory manner; but besides the two species already noticed, another, or a variety, is common; possibly,

Lepigonum salinum, *Fries.* Spergularia salina, Presl. (Gray's Manual.)
Growing in the salt marshes.

Spergula arvensis, *L.* (CORN SPURRY.)
Fields and roadsides. Quite common.

PARONYCHIEÆ.

(WHITLOW-WORT FAMILY.)

Anychia dichotoma, *Michx.* (FORKED CHICKWEED.)
A delicate inconspicuous plant, often overlooked, growing in dry places. Orne's point, Salem, 1824 (Dr. Charles Pickering); same locality, 1877 (J. R.); Boxford, 1878. Rather scarce.

Scleranthus annuus, *L.* (KNAWEL.)
A homely weed. Common near the coast. (Nat. from Europe.)

FICOIDEÆ.

Mollugo verticillata, *L.* (CARPET-WEED.)
Common in paths and in poor soil generally. (Int. from the south.)

PORTULACACEÆ.

(PURSLANE FAMILY.)

Portulaca oleracea, *L.* (COMMON PURSLANE.)
This omnipresent weed was considered an excellent table green by the early inhabitants, and so spoken of in letters written to Eng-

land. Some persons even consider this a native plant which is not the case. (Nat. from Eu.)

Claytonia Virginica, *L.* (SPRING BEAUTY.)
This plant is not represented in the county collection at the Peabody Academy of Science; and, although found abundantly north and west of us, is as yet only reported at "Haverhill, North Parish," (Mrs. Downs); and in "Lynnfield by Mr. Chute" (memo. G. D. Phippen).

MALVACEÆ.

(MALLOW FAMILY.)

Althæa officinalis, *L.* (COMMON MARSH-MALLOW.)
"Marsh, Salisbury, Sept." (Mrs. Downs). Not in collection at the Peabody Academy of Science. (Nat. from Eu.)

Malva rotundifolia, *L.* (COMMON MALLOW.)
A common door weed. (Nat. from Eu.)

Malva moschata, *L.* (MUSK MALLOW.)
Essex, Andover, etc. Escaped from old gardens. (Adv. from Eu.)

Abutilon Avicennæ, *Gærtn.* (VELVET-LEAF.)
Rubbish heaps, wharves, etc. Not rare. (Adv. from India.)

Hibiscus moscheutos, *L.* (SWAMP ROSE-MALLOW.)
"A variety of this or another species (palustris) grew in a ravine below Swampscott towards the beach," about 1824 (memo. Dr. Chas. Pickering). This locality is now occupied by residences. Cultivated by S. P. Fowler at Danvers, and by G. D. Phippen, Salem.

Hibiscus Trionum, *L.* (BLADDER KETMIA.)
An old garden flower occasionally escaping. Wenham (Mrs. M. W. Kimball), etc. Scarce. (Adv. from Eu.)

Hibiscus Syriacus, *L.* (SHRUBBY ALTHÆA.)
Much cultivated and escaping by seeds. (Adv. from Eu.)

TILIACEÆ.

(LINDEN FAMILY.)

Tilia Americana, *L.* (LINDEN; BASSWOOD.) Quite abundant particularly towards Boxford.

Tilia Europæa, *L.* (EUROPEAN LINDEN.)
Extensively planted as an ornamental shade tree, blooming nearly two weeks before the American tree and having much more fragrant blossoms.

LINACEÆ.

(FLAX FAMILY.)

Linum Virginianum, *L.*
"Very rare. Second pine hill, Lynn" (Tracy). Not represented in the county collection at the P. A. S.

Linum usitatissimum, *L.* (COMMON FLAX.)
Gloucester (Mrs. J. Babson); Salem (G. D. Phippen); Newburyport (J. R.); and occasionally in various towns.

GERANIACEÆ.

(GERANIUM FAMILY.)

Geranium maculatum, *L.* (WILD CRANESBILL.)
Common at the edges of woodlands.

Geranium Carolinianum, *L.*
Essex county, 1824 (Dr. Chas. Pickering); Lynn (Tracy); Salem Great Pastures; Danvers (J. H. Sears); Andover (Mrs. Downs). Not very common.

Geranium Robertianum, *L.* (HERB ROBERT.)
Stated by Josselyn (N. E. Rar., 1672) to have been brought by the colonists to New England and considered by Dr. Pickering (Chron. Hist. Pl., p. 168) a naturalized plant here. Common in damp rocky woods.

Erodium cicutarium, *L'Her.* (STORKSBILL.)
Banks of the Merrimac, near Lawrence, Sept. 1877 (J. R.); Saugus, May, 1880 (Mr. Smart). (Adv. from Eu.)

Impatiens fulva, *Nutt.* (WILD BALSAM.)
Common in damp shady places.

Oxalis Acetosella, *L.* (WOOD-SORREL.)
Danvers (Dr. Osgood's list). There is no specimen preserved of this plant and it may be that the O. stricta is intended, yet it is not at all improbable that this species should grow in many places in the county, although it has not been noticed by other botanists.

Oxalis violacea, *L.* (VIOLET WOOD-SORREL.)
Ipswich (Oakes). Rare.

Oxalis stricta, *L.* (YELLOW WOOD-SORREL.)
A common weed in gardens. A variety with deep red leaves is frequently met with.

RUTACEÆ.

(RUE FAMILY.)

Xanthoxylum Americanum, *Mill.*
Georgetown, introduced (Mrs. Horner); Boxford; Summer St. Cemetery, Salem; Topsfield; Danvers. Probably not a native of the county, but introduced from the north or west.

Ailanthus glandulosus, *Desf.* (TREE OF HEAVEN; CHINESE SUMACH.)
Often cultivated as an ornamental tree and freely produced from seeds. (Adv. from China.)

ANACARDIACEÆ.

(CASHEW FAMILY.)

Rhus typhina, *L.* (STAGHORN SUMACH.)
Common in rocky places and roadsides.
Rhus glabra, *L.* (SMOOTH SUMACH.)
Common. Particularly from Wenham to Topsfield and Beverly.
Rhus copallina, *L.* (DWARF SUMACH.)
Frequent with the last.
Rhus venenata, *DC.* (POISON SUMACH.)
Common in low ground and swamps.
Rhus Toxicodendron, *L.* (POISON IVY.)
Very common, both the low running, and the taller shrubby forms; the latter being most abundant near the sea.

VITACEÆ.

(VINE FAMILY.)

Vitis Labrusca, *L.* (NORTHERN FOX GRAPE.)
Common throughout the county. A variety with paler foliage and almost white fruit is found in Danvers by J. H. Sears.
Vitis æstivalis, *Michx.* (SUMMER GRAPE.)
Lynn (Tracy); Oakes Ledge, Danvers (Dr. Chas. Pickering); Cape Ann. Not so abundant as the last.
Vitis cordifolia, *Michx.* (FROST GRAPE.)
Essex, West Gloucester, Boxford, Methuen; Andover (Mrs. Downs), etc. Rather frequent.
Vitis heterophylla, *Thunb.* (CISSUS SIEBOLDII of the gardens.)
Found escaped at Nahant, 1878, by Miss Mary T. Saunders.

Ampelopsis quinquefolia, *Michx.* (VIRGINIAN CREEPER.)
Quite common in all parts of the county and a most valuable climber for cultivation.

RHAMNACEÆ.

(BUCKTHORN FAMILY.)

Rhamnus cathartica, *L.* (COMMON BUCKTHORN.)
In common use as a hedge plant and frequently seen as a small tree, often in quite out-of-the-way places. (Nat. from Eu.)

Ceanothus Americanus, *L.* (NEW JERSEY TEA.)
Beverly; Lynn (Tracy); Georgetown (Mrs. Horner); Bradford; Andover. Rather frequent.

CELASTRACEÆ.

(STAFF TREE FAMILY.)

Celastrus scandens, *L.* (ROXBURY WAX-WORK; CLIMBING BITTER-SWEET.)
Quite common. More frequent perhaps in the coast towns.

SAPINDACEÆ.

(SOAPBERRY FAMILY.)

Æsculus Hippocastanum, *L.* (HORSE-CHESTNUT.)
A common street-tree, springing up everywhere in towns. (Adv. from Asia.)

Acer Pennsylvanicum, L. (STRIPED MAPLE.)
Frequent in the older woods. It is very singular that its companion, A. spicatum, has never been found in the county.

Acer saccharinum, *Wang.* (ROCK MAPLE; SUGAR MAPLE.)
Common, although few large trees are to be found.
Var. **nigrum** is now and then met with under cultivation.

Acer dasycarpum, *Ehrh.* (WHITE MAPLE.)
Frequent along the Ipswich river at Topsfield, and in the N. W. portion of the county; also under cultivation as a shade tree.

Acer rubrum, *L.* (RED MAPLE; SWAMP MAPLE.)
Very common in moist soils. Very variable.

Acer Pseudo-Platanus and Acer platanoides, European trees, are frequent in cultivation, the latter producing many seedlings.

Negundo aceroides, *Mœnch.*
Has been found escaped, by seeds, in the vicinity of Boston. (In from Penn., etc.)

POLYGALACEÆ.

(MILKWORT FAMILY.)

Polygala sanguinea, *L.*
Common in damp places, often varying to white.

Polygala cruciata, *L.*
In bogs. Not very common. Marblehead (Tracy); Lynnfield (. P. Chute); Manchester (Oakes); Wenham swamp; Beaver por bog, etc.

Polygala verticillata, *L.*
Frequent in dry places.

Polygala polygama, *Walt.*
Wood paths and somewhat shady places. Quite common.

Polygala paucifolia, *Willd.* (FRINGED POLYGALA.)
Manchester, Essex; Georgetown (Mrs. Horner); Andover (Mr Downs), etc. Not very common.

LEGUMINOSÆ.

(PULSE FAMILY.)

Lupinus perennis, *L.* (WILD LUPINE.)
In the northern portion of the county from Ipswich to Andover, is quite common. "The flowers vary from white and pink to blu and purple, some striped" (memo. G. D. Phippen).

Genista tinctoria, *L.* (WOAD WAXEN.)
"By the first European colonists was carried to Salem in Ne England, 'woad-seed' being enumerated prior to February, 1628, i a memorandum of articles to be sent out with Governor Endicott forty years later, 'wood-wax wherewith they dye many prett colours" was found there by Josselyn (Rar. p. 51)." (Dr. Cha Pickering, Chron. Hist. Pl., p. 86).
"This plant has overrun the hills on the south side of Salem, s as to give them, in the month of July, a uniformly yellow appea ance at a distance" (Bigelow's Fl. Bost. 1814).
"In pastures between New-Mills and Salem" 1783 (Dr. Manasse Cutler, Mem. Am. Acad., Vol. I).
It now (1880) covers hundreds of acres of land on the sterile hil near Salem, Peabody and Danvers, extending somewhat into Middl ton and Topsfield. (Adv. from Eu.)

Trifolium arvense, *L.* (RABBIT-FOOT CLOVER.)
Common in dry soil. (Nat. from Eu.)
Trifolium pratense, *L.* (RED CLOVER.)
Everywhere in fields and by roadsides. (Adv. from Eu.)
Trifolium medium, *L.* (ZIGZAG CLOVER.)
Salem pastures, Danvers, and a few places in that vicinity. Not very common. (Adv. from Eu.)
Trifolium hybridum. (ALSYKE.)
Byfield, Georgetown, introduced and escaping. (Adv. from Eu.)
Trifolium repens, *L.* (WHITE CLOVER.)
Very common. "Probably introduced here but indigenous northward." (Gray's Manual.)
Trifolium agrarium, *L.* (YELLOW CLOVER.)
Roadsides. Scarce. (Nat. from Eu.)
Trifolium procumbens, *L.* (LOW YELLOW CLOVER.)
Danvers, Salem, Andover, etc. Quite common. (Nat. from Eu.)

Melilotus officinalis, *Willd.* (YELLOW MELILOT.)
Salem and Hamilton (G. D. Phippen); Salem Great Pastures, and occasionally elsewhere. (Adv. from Eu.)
Melilotus alba, *Lam.* (WHITE SWEET CLOVER.)
"Naturalized in Rowley" (Oakes in Hovey's Mag. Vol. XIII); "roadside near Beverly bridge" (G. D. Phippen); and other places. Frequent. (Adv. from Eu.)

Medicago lupulina, *L.* (BLACK MEDICK; NONESUCH.)
Roadsides everywhere. (Adv. from Eu.)
Medicago sativa, *L.* (LUCERNE.)
Escaped from cultivation. Danvers (J. H. Sears). Not well established. Danvers (Dr. Osgood's list), whether under cultivation or not is uncertain. (Adv. from Eu.)

Robinia Pseudacacia, *L.* (LOCUST.)
Thoroughly naturalized in some places. (Int. from Pennsylvania.)
Robinia viscosa, *Vent.* (CLAMMY LOCUST.)
Much cultivated and spreading. (Int. from Virginia.)
Robinia hispida, *L.* (ROSE ACACIA.)
Old gardens. (Int. from Virginia.)

Tephrosia Virginiana, *Pers.* (TEPHROSIA.)
In the N. W. portion of the county. Not very abundant. "Groveland in boggy land" (Proc. E. I., Vol. III, p. 18) seems improbable, as the plant grows in dry sandy soil.

Desmodium nudiflorum, *DC.*
Dry hilly woods. Common.

Desmodium acuminatum, *DC.*
Georgetown (Mrs. Horner); Andover (Mrs. Downs); Haverh (J. R.). Not very abundant.

Desmodium rotundifolium, *DC.*
Quite common in dry woods.

Desmodium cuspidatum, *Hook.*
"Dark Lane, Salem" (memo. Rev. J. L. Russell); "Andover (mem Mrs. Downs). Not represented in the county collection of tl P. A. S.

Desmodium paniculatum, *DC.*
"Oakes' ledge, Danvers, probably a different form from that in tl middle states" (memo. Dr. Chas. Pickering); Andover, etc.

Desmodium Canadense, *DC.*
Frequent in woodlands.

Desmodium Marilandicum, *Boott.*
In similar situations to the other species. The Desmodiums inhab a district from Ipswich, Hamilton, and Danvers, north and wes ward; absent in many places, and scarce in others outside of th: region. Tracy mentions only one species in the vicinity of Lyn:

Lespedeza violacea, *Pers.*
Dry woodlands. Frequent.

Lespedeza reticulata, *Pers.* Lespedeza *var.* violacea sessiliflo1 (Gray's Manual.)
Georgetown (Mrs. Horner); Danvers (J. H. Sears). Not rare.

Lespedeza hirta, *Ell.*
Dry places. Frequent.

Lespedeza capitata, *Michx.*
Common by roadsides.

Vicia sativa, *L.* (COMMON VETCH; TARE.)
Quite common. (Adv. from Eu.)

Vicia Cracca, *L.*
Georgetown (Mrs. Horner); West Newbury; Gloucester (Mrs. , Babson); Legges Hill, Salem (G. D. Phippen), etc. Considered b Dr. Chas. Pickering to be an introduced plant here. Scarce.

Lathyrus maritimus, *Bigelow.* (BEACH PEA.)
Common along the shore.

Lathyrus palustris, *L.* (MARSH VETCHLING.)
Ipswich (Oakes), "near Frye's mills, Salem 1824-5" (Dr. Cha. Pickering). Rather scarce.

Apios tuberosa, *Mœnch.* (GROUND NUT.)
Abundant in most parts of the county. An excellent climber fc cultivation the flowers being very fragrant.

Phaseolus diversifolius, *Pers.*
Near the coast. Not rare. A form with mostly undivided leaves and bearing tubers on the stem just below the ground; grows on Deer Island, in the Merrimac, at Newburyport.

Amphicarpæa monoica, *Ell.* (HOG PEANUT.)
Woodlands. Common.

Baptisia tinctoria, *R. Br.* (WILD INDIGO.)
Common in dry soil.

Cassia Marilandica, *L.* (WILD SENNA.)
"Near Salem" (Buttrick's list Proc. E. I., Vol. II, p. 24); Andover (G. D. Phippen); Georgetown (Mrs. Horner). Scarce.

Cassia nictitans, *L.* (WILD SENSITIVE-PLANT.)
Lawrence; Groveland (Rev. J. L. Russell); Deer Island, Newburyport; and a few other localities. Scarce.

Gleditschia triacanthos, *L.* (THREE-THORNED ACACIA; HONEY LOCUST.)
Introduced from Pennsylvania, etc., frequently spreading by seeds.

ROSACEÆ.

(ROSE FAMILY.)

Prunus Americana, *Marshall.* (WILD YELLOW OR RED PLUM.)
Represented in the P. A. S. county herbarium by a specimen collected by the late Mr. Oakes, the precise locality being unknown.

Prunus maritima, *Wang.* (BEACH PLUM.)
Near the coast. Common at Ipswich, Plum Island, etc. Found also in the interior along walls where the stones, thrown by farmers on their way home from haying on the marshes, have sprung up.

Prunus pumila, *L.* (DWARF CHERRY.)
"Outlet of mineral spring pond" (memo. Dr. Chas. Pickering); "Andover, June, 1873" (memo. Rev. H. P. Nichols). Not represented in the county collection at the P. A. S.

Prunus Pennsylvanica, *L.* (WILD RED CHERRY.)
Not uncommon in most parts of the county.

Prunus Virginiana, *L.* (CHOKE CHERRY.)
Very common near walls.

Prunus serotina, *Ehrhart.* (WILD BLACK CHERRY.)
Common. Often attaining large size. Considered by the late Dr. Chas. Pickering to have been introduced into the county by the early settlers, from other parts of New England.

Prunus domestica (PLUM),
Prunus Persica (PEACH), and
Prunus Cerasus (CHERRY), are of course frequent in cultivation; tl
two last named are often found by walls and roadsides escape
although seldom reaching any great size.

Spiræa salicifolia, *L.* (COMMON MEADOW-SWEET.)
Damp ground. Common.

Spiræa tomentosa, *L.* (HARDHACK.)
Very common in rather drier situations than the last.

Spiræa Ulmaria (Adv. from Eu.) and also the

Spiræa lobata (a western species), which are cultivated in o
gardens, are found escaped in Wenham, Topsfield and Danvers, t
the roadside. Scarce.

Spiræa sorbifolia, *L.*
Much cultivated; frequently runs wild; found escaped in Bever.
(Frank Stone). (Adv. from Eu.)

Poterium Canadense, *Benth & Hook.* (CANADIAN BURNET.)
Gloucester, scarce (Mrs. Babson); Hamilton; Ipswich; Topsfield
frequent in meadows; not found in the Salem or Lynn region.

Agrimonia Eupatoria, *L.* (COMMON AGRIMONY.)
Common in various parts of the county.

Geum album, *Gmelin.*
Very common by roadsides, etc.

Geum Virginianum, *L.*
Essex County (Dr. Chas. Pickering); "occasional in Lynn" (Tracy)
Not in the county collection at the P. A. S.

Geum strictum, *Ait.*
"Rare in Lynn" (Tracy); Wenham; Ipswich (Oakes); Danvei
(Buttrick); and in other places.

Geum rivale, *L.* (WATER OR PURPLE AVENS.)
Quite common in meadows.
Dr. Chas. Pickering found (1823–4) "a curious Geum at Orne
Point, Salem, with green petals; perhaps a variety of G. album, c
possibly another species."

Potentilla Norvegica, *L.*
Common in dry soil.

Potentilla Canadensis, *L.* (FIVE-FINGER.)
Hillsides everywhere. One of our earliest spring flowers.

Var. simplex, *T. & G.*
Later, but quite as common.

Potentilla argentea, *L.* (SILVERY CINQUE-FOIL.)
Railroad beds and roadsides. Common.

Potentilla arguta, *Pursh.*
Lynnfield Hotel Station (Mrs. Horner); Salem Pastures (S. B. Buttrick); Gloucester (Mrs. Babson); Andover (Mrs. Downs). Not very common.

Potentilla Anserina, *L.* (SILVER-WEED.)
Salt marshes. Common.

Potentilla fruticosa, *L.* (SHRUBBY CINQUE-FOIL.)
"Serpentine quarry, Lynnfield" (Tracy); Rockport (Frank Lufkin); "Turkey hill" and "Pine swamp," Ipswich. The stems of this species are sometimes more than half an inch in diameter. Although abundant in the above localities, this may be considered as scarce in the county. This plant is a great pest in portions of western Massachusetts.

Potentilla tridentata, *Ait.*
Gloucester (Mrs. Babson); Rockport (Frank Lufkin). Scarce.

Potentilla palustris, *Scop.* (MARSH FIVE-FINGER.)
"Danvers" (Dr. Osgood's list); Wenham (Dr. Chas. Pickering). Not in the county collection at the P. A. S. Scarce.

Fragaria Virginiana, *Ehrhart.* (COMMON STRAWBERRY.)
Damp ground or hillsides. Common.

Fragaria vesca, *L.* (LONG-FRUITED STRAWBERRY.)
Although this has been reported from several places, the only specimens seen were collected at Boxford. Scarce.

Rubus Dalibarda, *L.* (Dalibarda repens, *L.* Gray's Manual.)
Manchester woods (Mrs. Babson and others).

Rubus odoratus, *L.* (PURPLE FLOWERING-RASPBERRY.)
Amesbury (J. G. Whittier); Andover (Mrs. Downs), etc. Common in rocky places. Higginson speaks of this plant as growing near Salem in 1629, and the locality where it now flourishes in Salem Great Pastures was considered by Dr. Chas. Pickering to be the same one known to Higginson. Frequent in cultivation in the last century as noted by Cutler in 1783.

Rubus triflorus, *Richardson.* (DWARF RASPBERRY.)
"Essex County" (memo. Dr. Chas. Pickering); "Danvers" (Bigelow's Fl. Bost. 2nd. ed. 1824) 1877 (J. H. Sears). Growing in moist places rather than dry hills as spoken of by some writers. Scarce.

Rubus strigosus, *Michx.* (WILD RED RASPBERRY.)
Common by roadsides and in rocky places.

Rubus occidentalis, *L.* (THIMBLEBERRY.)
Frequent in most towns of the county.

Rubus Canadensis, *L.* (LOW BLACKBERRY; DEWBERRY.)
Fields and borders of woods. Common.

Rubus villosus, *Ait.* (HIGH BLACKBERRY.)
    The tall prickly form often growing ten feet high is common everywhere. The low form is also abundant. Another more bushy variety with abundant flowers, poor fruit, and very densely covered with short prickles, grows at Danvers (J. H. Sears). The smooth form which is found at the Profile House region in New Hampshire does not seem to grow here.
Rubus hispidus, *L.* (RUNNING SWAMP-BLACKBERRY.)
    Common, although hardly ever found in swamps.
Rosa Carolina, *L.* (SWAMP ROSE.)
    Abundant along moist roadsides.
Rosa lucida, *Ehrhart.* (WILD ROSE.)
    Very common. Varying somewhat according to situation.
Rosa rubiginosa, *L.* (SWEET-BRIER.)
    Common in fields and along the roadsides. (Nat. from Eu.)
Rosa micrantha, *Smith.* (SMALL-FLOWERED SWEET-BRIER.)
    "Danvers" (Dr. Osgood's list); "Cape Ann" (Mrs. Downs). Not in the county collection at the P. A. S. (Nat. from Eu.)
Rosa cinnamomea, cultivated in old gardens; often found in deserted places. (Adv. from Eu.)

Cratægus Oxyacantha, *L.* (ENGLISH HAWTHORN.)
    "Near Pranker's mills, Saugus" (Tracy); Derby estate, Salem. Cultivated and established in some places. (Adv. from Eu.)
Cratægus coccinea, *L.* (SCARLET-FRUITED THORN.)
    Ipswich, Topsfield, Danvers, etc. In the central and northern region more frequent than in the southern.
Cratægus tomentosa, *L.* (BLACK THORN.)
    In the region of the last. The form found in the county seems to be var. punctata; the leaves, however, vary considerably.

Pirus arbutifolia, *Ait.*, var. **melanocarpa.** (CHOKE-BERRY.)
    Common along roadsides in rocky places.
Var. **erythrocarpa.**
    Amesbury near the Merrimac shore.
Pirus Americana, *DC.* (AMERICAN MOUNTAIN-ASH.)
    Essex County (Oakes); "occasional at Lynn" (Tracy); Danvers (Dr. Osgood's list). Scarce.
Pirus aucuparia, *Gærtn.* (EUROPEAN MOUNTAIN-ASH.)
    Often cultivated, and as the fruit is much sought by birds, the young trees are frequently found escaped. (Adv. from Eu.)
Pirus malus, the Apple, and also
Pirus communis, the Pear, have been cultivated from the earliest settlement of the county. The former frequently, and the latter occasionally, are found escaped at long distances from cultivated lands.

Amelanchier Canadensis, *T. & G.* (SHADBUSH; JUNEBERRY.)
Common in low lands. This is the large form with long petals and reddish leaves in Spring. (Var. Botryapium Gray's Manual.)
Var. oblongifolia.
Equally common, but not so large as the last; leaves downy when young, petals shorter. Tracy speaks of "a curious variety found at "Norman's Woe," Gloucester, which fruits when only three feet high."

SAXIFRAGACEÆ.

(SAXIFRAGE FAMILY.)

Ribes oxycanthoides, *L.*, Ribes hirtellum, *Michx.* (Gray's Manual). (COMMON WILD GOOSEBERRY.)
Frequent in low grounds.

Ribes floridum, *L'Her.* (WILD BLACK CURRANT.)
Danvers, Amesbury, Ipswich (Oakes). Not rare. "Much like the Black Currant of the gardens" (Gray's Manual),

Ribes nigrum, of Europe, which is spontaneous near old gardens.

Ribes rubrum, *L.* (RED CURRANT.)
The specimens found in the county have probably all escaped from garden plants introduced from Europe, although this species is indigenous northward. Quite common in the older towns.

Ribes aureum, *Pursh.* (MISSOURI CURRANT.)
Common in old gardens and sometimes found in neglected places. (Int. from the West.)

Philadelphus coronarius, *L.* (MOCK ORANGE.)
This species is, in common with others of the genus, wrongly called "Syringa" which is the botanical name for the "Lilac." It is occasionally found by the roadside escaped from gardens. (Probably introduced from Japan.)

Parnassia Caroliniana, *Michx.* (GRASS OF PARNASSUS.)
Georgetown (Mrs. Horner); Lynnfield (Rev. A. P. Chute); "meadow near Howe's farm" (memo. Rev. J. L. Russell); Marblehead (Tracy); Hamilton (G. D. Phippen); Topsfield. Scarce, although found in a number of places.

Saxifraga Virginiensis, *Michx.* (EARLY SAXIFRAGE.)
Very common in exposed rocky places.

Var. chlorantha (Oakes, in Hovey's Mag. Vol. XIII) is a form having green flowers, probably an abnormal condition of the ordinary plant.

Saxifraga Pennsylvanica, *L.* (SWAMP SAXIFRAGE.)
Common in meadows.

Tiarella cordifolia, *L.* (FALSE MITRE-WORT.)
"Rare in Lynn, Dr. Holder legit" (Tracy). Only represented :
the P. A. S. herbarium by a cultivated specimen from the collectic
of Mr. G. D. Phippen.

Chrysosplenium Americanum, *Schwein.* (GOLDEN SAXIFRAGE
The name suggests a much more extensive and elegant plant tha
this humble greenish flowered species. Common in very wet placı
in or near woods.

CRASSULACEÆ.

(ORPINE FAMILY.)

Penthorum sedoides, *L.* (DITCH STONE-CROP.)
Common in wet places.

Sedum acre, *L.* (MOSSY STONE-CROP. GOLDEN MOSS.)
Beverly, Salem Great Pastures, Andover (G. D. Phippen), et
Frequent in exposed rocky places. (Adv. from Eu.)

Sedum ternatum, *Michx.*
Not a native but introduced into cultivation from the west. Tho
oughly escaped in Danvers by a roadside (J. R., 1877).

Sedum Telephium, *L.* (LIVE-FOR-EVER; AARON'S-ROD.)
In Wenham and Danvers, this has become one of the wor
weeds in grass land, and as every bit left to itself will soon tal
root, it is very difficult to exterminate. Escaped from old garden
Andover (Mrs. Downs); Lynn (Tracy). (Adv. from Eu.)

Sedum reflexum, *L.*
Pigeon Cove, escaped. (See Am. Nat., Sept., 1876.) (Adv. fro
Eu.)

Sempervivum tectorum, *L.* (HOUSELEEK.)
"Saugus and two places in Lynn" (Tracy); "for twenty or thir
years on rocks back of D. Nichols' house, Boston street, Salem
(memo. G. D. Phippen); Swampscott, on line of Eastern Railroa
well established (J. R.). (Adv. from Eu.)

HAMAMELACEÆ.

(WITCH-HAZEL FAMILY.)

Hamamelis Virginiana, *L.* (WITCH HAZEL.)
Quite common in most towns of the county.

HALORAGEÆ.

(Water-Milfoil Family.)

Myriophyllum spicatum, *L.*
Pleasant pond, Wenham; and some others.
Myriophyllum ambiguum, *Nutt.*
Var. **natans.**
"Breed's pond, Lynn" (Tracy).
Var. **capillaceum.**
"Rocky pond hole in Marshall's pasture, Lynn" (Tracy); Georgetown (J. H. Sears). Not rare.
Var. **limosum.**
Danvers, Dr. Nichols (Bigelow's Fl. Bost. 2d ed., 1824, under M. procumbens).
Myriophyllum tenellum, *Bigelow.*
Chebacco pond; Pleasant pond, Wenham; Wenham pond; "Danvers" (Dr. Osgood's list). Frequent.

Proserpinaca palustris, *L.* (Mermaid-weed.)
Not uncommon in wet places.

ONAGRACEÆ.

(Evening-Primrose Family.)

Circæa Lutetiana, *L.* (Enchanter's Night shade.)
Common in damp shady places.
Circæa alpina, *L.*
In similar places to the last, nearly as common.

Epilobium spicatum, *Lam.*, Epilobium angustifolium, *L.* (Gray's Manual.) (Willow-herb.)
Very common in gravelly soil, and particularly so in places recently burnt over; hence, often wrongly called "fire weed," the common name for Erechthites.
Epilobium palustre, *L.*, var. **lineare.**
Essex County (G. D. Phippen, Dr. Pickering). Not represented in the collection at the P. A. S.
Epilobium molle, *Torr.*
"Once in a pasture at Wenham, 1824" (letter from Dr. Charles Pickering). Not in the P. A. S. collection.
Epilobium coloratum, *Muhl.*
Very common in damp places and exceedingly variable in form.

Epilobium hirsutum, *L.*
Appeared in old gardens and waste heaps in Salem about 1860, and still continues in some places.

Œnothera biennis, *L.* (EVENING PRIMROSE.)
Very common and variable.

Var. **grandiflora** is often met with in cultivation, occasional in a wild state, its size is probably the result of rich soil, the flower often being four inches in diameter.

Var. **parviflora.** In gravelly soil.

Var. **cruciata.** "E. R. R. road-bed in Wenham or North Beverly 1870" (G. D. Phippen).

Œnothera fruticosa, *L.* (SUNDROPS.)
Beverly Farms (John C. Phillips). Perhaps introduced. A beautiful plant for cultivation.

Œnothera pumila, *L.*
Common. About as variable as Œ. biennis. Delicate specimens only a few inches high, and stout ones three feet tall, have been found; the latter may have been from the biennial roots spoken of in Gray's Manual.

Ludwigia alternifolia, *L.* (SEED-BOX.)
North Andover (Russell); Amesbury (J. R.); Lynn (Tracy). Rather scarce.

Ludwigia palustris, *L.* (WATER PURSLANE.)
Muddy places. Quite common.

MELASTOMACEÆ.

(MELASTOMA FAMILY.)

Rhexia Virginica, *L.* (MEADOW-BEAUTY.)
Borders of ponds and swampy places. Not very common except in the central and northern portions of the county.

LYTHRACEÆ.

(LOOSESTRIFE FAMILY.)

Ammannia humilis, *Michx.*
"Danvers, 1818, Dr. Nichols (Oakes in Hovey's Mag. Vol. XIII) "muddy strand and dry pools, Humphrey's pond" (memo. Dr. Chas Pickering).

Lythrum Hyssopifolia, *L.*
Common near the coast. Varying from small to very large forms, some specimens being twenty-six inches high and much branched. A few specimens found at Boxford, Aug. 1880, twelve miles inland.

Lythrum Salicaria, *L.* (SPIKED LOOSESTRIFE.)
Georgetown (Mrs. Horner). Scarce. Dr. Chas. Pickering considered both Lythrums to be introduced species, at least in Essex county.

Nesæa verticillata, *H. B. K.* (SWAMP LOOSESTRIFE.)
Common in wet places, and borders of ponds and streams. The portions of the long branches which droop over into the water become much enlarged by a corky outside tissue.

CACTACEÆ.

(CACTUS FAMILY.)

Opuntia vulgaris, *Mill.* (PRICKLY PEAR.)
Near "Kernwood" in Salem, some thirty years ago (Hugh Wilson); also on the Ipswich river bank at North Reading, beyond Middleton; where a few plants were placed many years ago there is now a very flourishing locality. The natural habitat of this species is from Nantucket southward.

CUCURBITACEÆ.

(GOURD FAMILY.)

Sicyos angulatus, *L.* (STAR CUCUMBER.)
A weed in waste places.

Echinocystis lobata, *Torr & Gray.*
Cultivated, and often escaped. Probably introduced from farther west.

UMBELLIFERÆ.

(PARSLEY FAMILY.)

Hydrocotyle Americana, *L.*
Common in damp wood-paths.

Hydrocotyle umbellata, *L.*
Chebacco and some other ponds, not flowering abundantly; usually found growing under water, flowering as the water recedes.

Sanicula Marilandica, *L.* (BLACK SNAKE-ROOT.)
Frequent throughout the county.

Daucus Carota, *L.* (COMMON CARROT.)
Common in fields and along the roadsides. (Nat. from Eu.)

Heracleum lanatum, *Michx.* (COW-PARSNIP.)
In most of the towns, near walls. Not very common.

Pastinaca sativa, *L.* (COMMON PARSNIP.)
Common. Fields and roadsides. (Adv. from Eu.)

Archangelica atropurpurea, *Hoffm.* (GREAT ANGELICA.)
North Andover and Wenham swamp (Dr. Chas. Pickering); Top field, Peabody, West Newbury (Wm. Merrill). Rather scarce.

Archangelica Gmelini, *DC.*
"Essex Co. (Dr. Pickering); and also along the coast" (Oakes Hovey's Mag. Vol. XII); "Topsfield and Scituate—Mr. Oake Mr. Russell." (Bigelow's Fl. Bost., 3d ed. 1840, under Ligusticur actæifolium); "Kernwood, Salem, 1863" (memo. Rev. J. L. Russell) Magnolia (C. E. Faxon). Not rare near the coast.

Æthusa Cynapium, *L.* (FOOL'S PARSLEY.)
Gloucester (C. E. Faxon); "occasional in Lynn" (Tracy); Ipswic (Oakes). Rare.

Carum Carui, *L.* (GARDEN CARAWAY.)
"Naturalized at Rowley and Ipswich" (Oakes in Hovey's Ma; Vol. XIII); Wenham (J. R.). Scarce. (Adv. from Eu.)

Levisticum officinale, *L.* (LOVAGE.)
Escaped from old gardens. Wenham, 1875, etc. Not common. (Ad· from Eu.)

Ligusticum Scoticum, *L.* (SCOTCH LOVAGE.)
Lynn (Tracy); islands in Salem harbor and along the coast. Cor mon.

Thaspium aureum, *Nutt.* (MEADOW-PARSNIP.)
Andover (Rev. H. P. Nichols); Methuen (J. R.); Georgetow (Mrs. Horner); meadows and wet places. Common only in tl northwestern portion of the county, absent elsewhere.

Cicuta maculata, *L.* (SPOTTED COWBANE.)
Wet places. Common.

Cicuta bulbifera, *L.*
Common in meadows by the sides of brooks.

Sium cicutæfolium, *Gmelin.* Sium lineare, *Michx.* (Gray's Ma ual.) (WATER PARSNIP.)
Common in wet places.

Cryptotænia Canadensis, *DC.* (HONEYWORT.)
Ipswich (Oakes); Boxford; Andover. Frequent in hilly copses.

Osmorrhiza longistylis, *DC.* (SMOOTHER SWEET CICELY.)
"Paradise," Salem, 1824 (Dr. Chas. Pickering); Andover (Rev. H. P. Nichols); Swampscott (J. R.). Occasional.

Osmorrhiza brevistylis, *DC.* (HAIRY SWEET CICELY.)
Hamilton, Swampscott, Haverhill, Georgetown, etc. Neither species is very common.

Conium maculatum, *L.* (POISON HEMLOCK.)
Waste places. Common. Mr. G. D. Phippen mentions that prior to 1840 the present site of Lynde Block, Salem, was covered with this plant. In 1864, the whole area was burnt over, but in 1878 after the soil in the yard of the Museum building now adjoining, and then a portion of that lot, had been upturned, the Conium again appeared. (Nat. from Eu.)

ARALIACEÆ.

(GINSENG FAMILY.)

Aralia racemosa, *L.* (SPIKENARD.)
Haverhill (Mrs. Downs); North Andover (Rev. J. L. Russell); "Dungeon Rock, Lynn" (Tracy); Chebacco, Essex woods, Bradford (J. R.). Rather scarce.

Aralia hispida, *Vent.* (BRISTLY SARSAPARILLA.)
Cleared rocky places. Common.

Aralia nudicaulis, *L.* (WILD SARSAPARILLA.)
Dry woods. Common.

Aralia trifolia, *Gray.* (DWARF GINSENG.)
Common in the older woods. Formerly at Orne's Point, Salem (Dr. Chas. Pickering).

Hedera Helix, the European Ivy, is not sufficiently hardy to become spontaneous, although frequently planted out as a climber. Only noticed in flower when unprotected in Salem (Dr. Fiske), Sept., 1880.

CORNACEÆ.

(DOGWOOD FAMILY.)

Cornus Canadensis, *L.* (DWARF CORNEL; BUNCH-BERRY.)
Common in damp woods.

Cornus florida, *L.* (FLOWERING DOGWOOD.)
Essex, Gloucester; Pirate's Glen, Lynn (Tracy); Wenham swamp islands, 1824 (Dr. Chas. Pickering); Boxford and towns in that region (J. H. Sears). Not very common. The Boxford locality is about the northern limit of this species.

Cornus circinata, *L'Her.* (ROUND-LEAVED CORNEL.)
In most towns, but not very common.

Cornus sericea, *L.* (SILKY CORNEL.)
In wet places. Frequent.

Cornus stolonifera, *Michx.* (RED-OSIER DOGWOOD.)
Lynn (Tracy); Georgetown (Mrs. Horner); Wenham, Topsfield (Oakes). Not very common.

Cornus paniculata, *L'Her.*
Roadsides, etc. Common.

Cornus alternifolia, *L.*
Sometimes forming a good sized tree, although generally a shrub by roadsides. Common.

Nyssa multiflora, *Wang.* (TUPELO.)
A striking tree in wet or even dry soil. Common.

CAPRIFOLIACEÆ.

(HONEYSUCKLE FAMILY.)

Linnæa borealis, *Gronov.* (LINNÆA.)
Not uncommon in many parts of the county, in the older woods.

Symphoricarpus racemosus, *Michx.* (SNOWBERRY.)
A native from farther north and west. Common in cultivation and escaped in some places.

Lonicera sempervirens, *Ait.* (TRUMPET HONEYSUCKLE.)
Ipswich (Mrs. M. W. Kimball); "rocky cliff in Marblehead" (Rev. J. L. Russell, Proc. E. I., Vol. I, p. 273); "stone wall in Salem Great Pastures, 1866" (memo. G. D. Phippen); Topsfield (J. H. Sears). If not a native of the county, this plant has become one to all appearances. Gray mentions no nearer natural habitat than New York whence it has been introduced.

Var. **flava.** (GARDEN YELLOW HONEYSUCKLE.)
In the woods, Beverly, 1879 (Wm. G. Barton).

Lonicera ciliata, *Muhl.* (FLY-HONEYSUCKLE.)
Rockport (C. W. Pool); Beverly, Manchester (Oakes); Haverhill (Mrs. Downs); etc. Not very common.

Diervilla trifida, *Mœnch.* (BUSH-HONEYSUCKLE.)
Common. Banks and roadsides. A variety grows in Essex with smoother leaves and with but one flower in each axil.

Triosteum perfoliatum, *L.* (TRIOSTEUM.)
Lynn (Tracy); Ipswich (Oakes); Boxford, Haverhill, etc. Not very common.

Sambucus Canadensis, *L.* (COMMON ELDER.)
Frequent by roadsides in damp places.

Sambucus pubens, *Michx.* (RED-BERRIED ELDER.)
Towns on the Merrimac; Rockport; Essex; "Salem, 1840" (memo. Rev. J. L. Russell). In many places but not so abundant as the last.

Viburnum Lentago, *L.* (SHEEP-BERRY.)
Very common.

Viburnum nudum, *L.* (WITHE-ROD.)
Danvers (S. P. Fowler); Essex, Haverhill, Amesbury, etc. Not so common as the last.

Viburnum dentatum, *L.* (ARROW-WOOD.)
Frequent by roadsides.

Viburnum acerifolium. (DOCKMACKIE.)
Often met with in the older woods.

Viburnum Opulus, (CRANBERRY TREE.)
Wenham swamp, 1877 (J. H. Sears); Orne's Point, Salem, 1878 (J. R., perhaps introduced). The Wenham swamp locality is without doubt a natural one. Frequent in cultivation.

Viburnum lantanoides, *Michx.* (HOBBLE-BUSH.)
Skirting old woods. In many towns, yet not very common.

RUBIACEÆ.

(MADDER FAMILY.)

Galium asprellum, *Michx.* (ROUGH BEDSTRAW.)
Common in swampy places climbing over bushes.

Galium trifidum, *L.*, var. **tinctorium.** (SMALL BEDSTRAW.)
Common in swamps.

Var. ———?, a very pretty plant with many flowers. Danvers, 1878 (J. H. Sears).

Galium triflorum, *Michx.* (SWEET-SCENTED BEDSTRAW.)
"Frequent among ferns," Lynn (Tracy); Essex County, 1824 (Dr. Chas. Pickering). In many other places, but not so abundant as the last.

Galium circæzans, *Michx.* (WILD LIQUORICE.)
Chebacco, Newbury, Amesbury, Boxford, etc. Not uncommon in the woods.

Galium lanceolatum, *Torrey.*
Chebacco, Boxford, etc. Not uncommon in similar places to the last.

Cephalanthus occidentalis, *L.* (BUTTON-BUSH.)
Around pond holes and brooks. A homely shrub and very common.

Mitchella repens, *L.* (MITCHELLA.)
Very common, forming carpets in the woods. Rarely found with white berries.

Houstonia cærulea, *L.* (HOUSTONIA.)
Very common in pastures and by roadsides. A most lovely spring flower varying from white to quite deep blue.

DIPSACEÆ.

(TEASEL FAMILY.)

Dipsacus sylvestris, *Mill.* (WILD TEASEL.)
Danvers, 1853 (Dr. Osgood's list) and noticed later by Mr. Sears.

COMPOSITÆ.

(COMPOSITE FAMILY.)

Liatris scariosa, *Willd.* (BLAZING-STAR.)
Abundant in the region of Topsfield; "scarce in Lynn" (Tracy); Salisbury (G. D. Phippen); "Howes farm, Danvers, 1829" (memo. Rev. J. L. Russell). Not observed in the Cape region.

Eupatorium purpureum, *L.*
Very common in damp soil. "Eight feet high in Middleton, Aug. 21, 1861." (Proc. E. I., Vol. III).

Eupatorium teucrifolium, *Willd.*
Georgetown; Essex; Lynnfield (Rev. A. P. Chute); "Bowler swamp, Lynn" (Tracy); Hamilton (G. D. Phippen); Danvers (Dr. Osgood's list). Scarce.

Eupatorium rotundifolium, *L.*
In Middlesex near Essex county (Geo. E. Davenport).
Eupatorium pubescens, *Muhl.*
"Rare. Stone barn, Swampscott, and Johnson's swamp" (Tracy); Essex (J. R.); Ipswich (Oakes); Topsfield (J. H. Sears). Scarce.
Eupatorium perfoliatum, *L.* (THOROUGHWORT.)
Meadows. Common.
Eupatorium sessilifolium, *L.* (UPLAND BONESET.)
"Rare. Burrill's Hill, Lynn" (Tracy); Andover (Mrs. Downs); etc. Scarce.
Eupatorium ageratoides, *L.* (WHITE SNAKE-ROOT.)
Ipswich (Oakes); Georgetown (Mrs. Horner). Scarce.

Mikania scandens, *L.* (CLIMBING HEMP-WEED.)
West Newbury (W. P. Conant); Saugus; Lynn; Wenham Swamp; Hamilton (G. D. Phippen); Georgetown (Mrs. Horner). Frequent.

Tussilago Farfara, *L.* (COLTSFOOT.)
Baldwin lot, Webb St., Salem (G. D. Phippen); Georgetown (Mrs. Horner); Manchester, Ipswich, etc. Scarce. (Nat. from Eu.)

Sericocarpus solidagineus, *Nees.*
Frequent. Ipswich (Oakes); Swampscott; etc.
Sericocarpus conyzoides, *Nees.*
Among bushes. More common than the last.

Aster corymbosus, *Ait.*
Quite common in wooded places.
Aster macrophyllus, *L.*
Ipswich (Oakes); Boxford; Kernwood, Salem (J. R.); etc. Scarce. The radical leaves conspicuous in the woods where this species grows.
Aster patens, *Ait.*
Dry places. Common.
Aster lævis, *L.*, Var. **lævigatus.**
Dry woods. Common.
Var. **cyaneus.**
Similar places.
Aster undulatus, *L.*
Dry places. Abundant.
Aster cordifolius, *L.*
Common in dry places.
Aster ericoides, *L.*
Essex County (Oakes?) in herb. P. A. S; "Andover" (memo. Mrs. Downs). Rare.

Aster multiflorus, *Ait.*
Roadsides. Common.
Aster dumosus, *L.*
Borders of woods. Common.
Aster Tradescanti, *L.*
Roadsides. Common.
Aster miser, *L., Ait.*
Pastures and roadsides. Common.
Aster simplex, *Willd.*
"Lynn occasionally" (Tracy); Andover (Mrs. Downs). Shore of the Merrimac at Lawrence and Bradford. Not very common.
Aster carneus, *Nees.*
Ipswich (Oakes); Beverly (J. R.); etc. Not very common.
Aster longifolius, *Lam.*
Common in wet places.
Aster puniceus, *L.*
Moist places. Common.
Aster amethystinus, *Nutt.*
"Salem" (Gray's Manual, p. 234). Not in the P. A. S. herbarium.
Aster Novæ-Angliæ, *L.*
Banks, and along streams and walls. Common.
Aster acuminatus, *Michx*
In quite deep woods. Frequent.
Aster nemoralis, *Ait.*
"Bogs, Essex county" (memo. Dr. Chas. Pickering). A variety quite like the mountain form spoken of in Gray's Manual; grows at Crooked pond, Boxford.
Aster linifolius, *L.*
Salt marshes. Common.
Aster flexuosus, *Nutt.*
Is found beyond the county line southward.
Aster linariifolius, *L.,* Diplopappus linariifolius, *Hook.* (Gray's Manual).
Dry fields. Common.
Aster umbellatus, *Mill.,* Diplopappus umbellatus, *Torr & Gray* (Gray's Manual).
By walls and in damp places. Common.
Aster infirmus, *Michx.,* Diplopappus cornifolius, *Darl.* (Gray's Manual).
"Essex County" (memo. Dr. Chas. Pickering).

Erigeron Canadensis, *L.* (HORSE-WEED.)
Very common in cultivated land and by the roadsides. A native appearing like an introduced plant.

Erigeron bellidifolius, *Muhl.* (ROBIN'S PLANTAIN.)
Open ground. Not rare.

Erigeron Philadelphicus, *L.* (COMMON FLEABANE.)
Essex county (memo. G. D. Phippen); Danvers, 1853 (Dr. Osgood's list). No authentic specimens of this species have been seen, and it is doubtful as an Essex county plant.

Erigeron annuus, *Pers.* (DAISY FLEABANE.)
Fields and roadsides. Common.

Erigeron strigosus, *Muhl.*
In similar situations to the last. Common.

Solidago squarrosa, *Muhl.*
In a letter to Dr. Chas. Pickering, Feb. 20, 1828, Wm. Oakes, speaking of what he had found new while botanizing, says, "I found a fine spot of S. squarrosa in your locality;" which Dr. Pickering said must have been Essex Co. The only specimens yet collected are from Beverly, 1879 (J. H. Sears). This species is more common in the mountain regions of Maine, New Hampshire and Pennsylvania.

Solidago bicolor, *L.* (PALE GOLDEN-ROD.)
Dry places. Common.

Solidago latifolia, *L.*
Georgetown, Boxford, Amesbury, etc. Not very common.

Solidago cæsia, *L.*
Moist woods. Frequent.

Solidago puberula, *Nutt.*
Common near the coast, less so in the interior towns.

Solidago stricta, *Ait.*
Wenham swamp, 1824 (Dr. Chas. Pickering); "Bowler swamp and similar localities in Lynn" (Tracy). Not in the P. A. S. county herbarium. It is very possible there is some mistake and that this species does not grow here, as no specimens have since been found and forms of S. neglecta are sometimes mistaken for it.

Solidago speciosa, *Nutt.*
Ipswich (Oakes); Georgetown (Mrs. Horner); Danvers (J. H. Sears). Not uncommon.

Solidago sempervirens, *L.* (SEASIDE GOLDEN-ROD.)
A very handsome species along the ocean shore. Common.

Solidago elliptica, *Ait.*
This species probably comes within our county flora as it has been found south of the county line. A species found at Point Shirley (C. E. Faxon), supposed to be this, now proves to be a hybrid between S. sempervirens and some other species.

Solidago neglecta, *Torr & Gray.*
Swamps. Common. A large form grows at Essex with very exextensive panicles of flowers.

Solidago arguta, *Ait.*, var. **juncea.**
Abundant along walls and in rather moist places. It is a remarkably handsome species.

Solidago Muhlenbergii, *Torr. & Gray.*
Georgetown (Mrs. Horner); Boxford; Salem (Rev. J. L. Russell); Haverhill, etc. Not very common, and mostly in the older woods.

Solidago linoides, *Solander.*
Andover (Mrs. Downs); "occasional in Lynn" (Tracy). Not in the P. A. S. county herbarium.

Solidago altissima, *L.*
Common along the roadsides by walls.

Solidago odora, *Ait.* (SWEET GOLDEN-ROD.)
Beverly, Hamilton, Ipswich (Oakes), Amesbury, Methuen, etc. Not very common.

Solidago nemoralis, *Ait.*
A small but gorgeous species with one-sided racemes. Common in dry fields.

Solidago Canadensis, *L.*
Common by the roadsides, with S. altissima.

Solidago serotina, *Ait.*
Resembles the last. Not so common. "Occasional in Lynn" (Tracy), Beverly, Wenham, etc. Abundant at Revere station (C. E. Faxon).

Solidago lanceolata, *L.*
Common along the roadsides and by walls. Most of the Goldenrods are quite interesting under cultivation; many have been so treated with success by Mr. G. D. Phippen, of Salem.

Inula Helenium, *L.* (COMMON ELECAMPANE.)
"Rare in Lynn" (Tracy); Hamilton (G. D. Phippen); North Andover, Topsfield, Lawrence, etc. Frequent along the roadsides and near old houses in the central county towns. (Nat. from En.)

Inula salicina, *W.*
Escaped at Tapleyville, near the old carpet factory (J. H. Sears). Undoubtedly introduced with wool. (Adv. from Eu.)

Pluchea camphorata, *DC.* (SALT-MARSH FLEABANE.)
Common on all the salt marshes.

Silphium perfoliatum, *L.* (CUP-PLANT.)
A western species escaped in Danvers (J. H. Sears).

Iva frutescens, *L.* (HIGHWATER SHRUB.)
Frequent along the edges of the salt marshes.

Ambrosia artemisiæfolia, *L.* (ROMAN WORMWOOD.)
A weed. The cutting of the leaves very variable.

Xanthium strumarium, *L.*, var. **echinatum.** (COMMON SEA-COCKLEBUR.)
Beaches and waste places near the coast, sometimes in the interior, as near the dam at Lawrence and the carpet factory at Tapleyville.

Xanthium spinosum, *L.* (SPRING CLOTBUR.)
Near Flax pond, Lynn (Herbert A. Young, C. E. Faxon). "Introduced into Essex county as late perhaps as 1814, earlier authors not mentioning it" (Dr. Chas. Pickering, Chron. Hist. Pl. p. 976). (Nat. from Trop. Amer.).

Rudbeckia laciniata, *L.*
Frequent. Georgetown; "Dark lane, Salem, 1827" (memo. Rev. J. L. Russell); Beverly, etc. "Abundant at Hamilton near Ipswich;" "a noble plant in cultivation" (G. D. Phippen).

Rudbeckia hirta, *L.* (CONE-FLOWER.)
Introduced from the west and fast becoming common in fields where it is likely to become troublesome. In a letter written July 15, 1875, the late Dr. Chas. Pickering says "R. hirta has entered Essex county since I left in 1826. It was observed in Pennsylvania by Muhlenberg, by myself near Philadelphia, and is entered on my Catalogue (finished in 1837) as not north of lat. 40°. How far the invasion is attributable to the removal of the forest remains an open question.'

Helianthus annuus, *L.* (GARDEN SUNFLOWER.)
Very common in cultivation, and occasionally spontaneous in waste places a few seasons. (Int. from Trop. Am.)

Helianthus lenticularis. (A form of H. annuus).
Two feet high, flowers small ($1'-3'$ in diameter) centre black. Tapleyville (J. H. Sears). Introduced with wool. A "double sunflower" has also been growing for nearly thirty years in a field in Danvers (J. H. Sears).

Helianthus strumosus, *L.*
Quite common. In damper places than the next.

Helianthus divaricatus, *L.*
Common in dry thinly wooded places and along roadsides and walls.

Helianthus tuberosus, *L.* (JERUSALEM ARTICHOKE.)
Escaped from old gardens along the roadsides. (Adv. from Eu.)

Coreopsis tinctoria, *Nutt.*
Often cultivated from the west in gardens. Found growing spontaneously in a field in Wenham, 1876 (Mrs. M. W. Kimball).

Coreopsis trichosperma, *Michx.* (TICKSEED SUNFLOWER.)
"Occasional in Lynn" (Tracy). Very common in many other towns. The paths in Wenham swamp in August are golden with the flowers of this species which often grows to a height of two or three feet, while in other places the plants are but a few inches high.

Bidens frondosa, *L.* (COMMON BEGGAR-TICKS.)
   A common weed.
Bidens connata, *Muhl.* (SWAMP BEGGAR-TICKS.)
   Frequent. West Newbury (W. P. Conant); Lynn (Tracy); etc. Forms of this species are sometimes mistaken for B. cernua.
Bidens chrysanthemoides, *Michx.* (LARGER BUR-MARIGOLD.)
   Common in swamps. (B. cernua in Tracy's Lynn Flora.)
Bidens Beckii, *Torrey.* (WATER MARIGOLD.)
   Hamilton, 1857 (G. D. Phippen); Ipswich river; Crane pond, West Newbury (J. R.); and occasionally elsewhere. Scarce.
Helenium autumnale, *L.* (SNEEZE-WEED.)
   Danvers (J. H. Sears). A western species of recent introduction.
Maruta Cotula, *DC.* (COMMON MAY-WEED.)
   Very common by roadsides, and in yards in the country towns. The English name does not apply here, as it does not flower until July or the last of June. (Nat. from Eu.)
Achillea Millefolium, *L.* (YARROW.)
   Common in fields, etc. The pink variety is found in Georgetown (Mrs. Horner); Lowell Island, Salem Harbor (G. D. Phippen); and now and then in other places.
Achillea Ptarmica, *L.* (SNEEZEWORT.)
   "Danvers, Mr. Oakes" (Bigelow's Fl. Bost. 3d ed. 1840). "Danvers locality thought to be destroyed" (memo. J. H. Sears 1875); "occasional in Lynn" (Tracy); Salem (Rev. J. L. Russell); "Lynn woods" (memo. G. D. Phippen). Scarce. (Adv. from Eu.)
Leucanthemum vulgare, *Lam.* (WHITEWEED.)
   "Fields and meadows: too abundant" (Gray's Manual). "Very injurious to grass lands" (Rev. M. Cutler, 1783, Proc. Am. Acad. Vol. I). A very curious monstrosity of this very common species was found at Wenham by Laurence G. Kemble having twelve smaller heads arising from the first one, thus beautifully illustrating the construction of a composite flower. (Nat. from Eu.)
Leucanthemum Parthenium, *Godron.* (FEVERFEW.)
   Common in old gardens and occasionally escaping as in Byfield, North Andover, etc. (Adv. from Eu.)
Pyrethrum Tanacetum, *DC.* (Matricaria Tanacetum, Wood's Bot. and Fl., M. Balsamita, Wood's Cl. Bk.) Misery Island, Salem Harbor, near an old farm. (Adv. from Eu.)
Tanacetum vulgare, *L.* (COMMON TANSY.)
   Roadsides. Very common. (Nat. from Eu.)
Var. **crispum.**
   Beverly and elsewhere. Not so common. (Nat. from Eu.)

Artemisia caudata, *Michx.*
Along the shore. Common. Artemisia Canadensis, *Michx.*, from Plum Island, in Bigelow's Fl. Bost. 3d ed., 1840, is evidently this species.

Artemisia vulgaris, *L.* (COMMON MUGWORT.)
Amesbury, Beverly, etc. Not very common.

Artemisia biennis, *Willd.*
Spoken of by Gray as "rapidly extending eastward by railroad;" has appeared in the filled land by the N. Y. & N. E. R. R. freight station in Boston (Sept., 1877), and probably is to be found elsewhere in this vicinity.

Artemisia Absinthium, *L.* (COMMON WORMWOOD.)
Salem, and a few other towns, escaped in fields. (Adv. from Eu.)

Artemisia Stellerianum, *Bess.* (One of the Dusty Millers.)
Nahant Beach (1877) northern end, growing in sand, and (1880) evidently increasing quite rapidly. (Adv. from Eu.)

Gnaphalium decurrens, *Ives.* (EVERLASTING.)
Rare in Lynn (Tracy); Danvers (Peabody?) (Rev. J. L. Russell); Andover (Mrs. Downs). Scarce.

Gnaphalium polycephalum, *Michx.* (COMMON EVERLASTING.)
Fields. Common.

Gnaphalium uliginosum, *L.* (LOW CUDWEED.)
Cart-paths and sterile fields. Very common.

Antennaria margaritacea, *R. Brown.* (PEARLY EVERLASTING.)
Common in dry fields.

Antennaria plantaginifolia, *Hook.*
One of our earliest flowers. Very Common.

Erechthites hieracifolia, *Raf.* (FIREWEED.)
On land recently burned over. Very common.

Senecio vulgaris, *L.* (COMMON GROUNDSEL.)
A weed. (Nat. from Eu.)

Senecio aureus, *L.* (GOLDEN RAGWORT.)
Swamps. Common.

Var. **obovatus.**
Hillside at Boxford; Danvers, 1826 (Dr. Nichols.)

Var. **Balsamitæ.**
Ipswich (Oakes); Swampscott (G. D. Phippen), etc. Common.

Centaurea Cyanus, *L.* (BLUEBOTTLE.)
Old gardens, Beverly (Frank Stone), etc., escaping for a few years. (Adv. from Eu.)

Centaurea nigra, *L.* (KNAPWEED.)
"One specimen on railroad roadbed Swampscott" (Tracy); Danvers (Oakes); "Dark Lane, Salem" (memo. G. D. Phippen), now extinct; Marblehead (G. D. Phippen); Topsfield. Not very common except in the last named locality. (Adv. from Eu.)

Cirsium lanceolatum, *Scop.* (COMMON THISTLE.)
Common. Roadsides, etc. (Nat. from Eu.)

Cirsium discolor, *Spreng.*
"Rare in Lynn" (Tracy); near Salem or Wenham, 1824 (Dr. Chas. Pickering); Danvers, etc. Not a very common species.

Cirsium muticum, *Michx.* (SWAMP THISTLE.)
Wenham swamp, 1824-6 (Dr. Chas. Pickering); Danvers, Methuen, Amesbury, etc. Frequent in low grounds.

Cirsium pumilum, *Spreng.* (PASTURE THISTLE.)
Fields, etc. Common. A fragrant and handsome flower.

Cirsium horridulum, *Michx.* (YELLOW THISTLE.)
Cape Ann and about Salem, 1824 (Dr. Chas. Pickering); Manchester (G. D. Phippen); Pigeon Cove; Plum Island. Rather scarce.

Cirsium arvense, *Scop.* (CANADA THISTLE.)
A very troublesome weed. (Nat. from Eu.)

Onopordon acanthium, *L.* (COTTON THISTLE.)
Little Nahant, Lynn, Salem Neck. Generally near the shore but seen in Boxford, perhaps introduced with rock-weed manure from some beach. (Adv. from Eu.)

Lappa officinalis, *Allioni.* (BURDOCK.)
Common everywhere. (Nat. from Eu.)

Lampsana communis, *L.* (NIPPLE-WORT.)
Salem, 1857 (Rev. J. L. Russell); Norman St., Salem, 1877 (J. R.). Rare. (Adv. from Eu.)

Cichorium Intybus, *L.* (CICHORY.)
"Fields in Cambridge," 1783 (Dr. Manasseh Cutler in Mem. Am. Acad., Vol. I). This would seem to indicate that at this time it had not reached Essex Co. Bigelow (1814) speaks of it as "everywhere in the vicinity of Boston." Very abundant on the roadbed of the Eastern railroad at Somerville and Lynn; Salem near March St., etc.; Gloucester (Mrs. Babson); Ipswich, Newburyport. Increasing. (Nat. from Eu.)

Krigia Virginica, *Willd.* (DWARF DANDELION.)
Common on dry hills. Flowers a beautiful orange color, heads small.

Leontodon autumnale, *L.* (FALL DANDELION.)
Very common. Flowering as early as June 15th in 1874. (Nat. from Eu.)

Hieracium Canadense, *Michx.*
Topsfield, Manchester, Newbury, Salisbury, Lynn, etc. Common.

Hieracium scabrum, *Torrey.*
Common in dry woods or on banks.

Hieracium venosum, *L.* (RATTLESNAKE-WEED.)
Dry rocky woods. Rather scarce.

Hieracium paniculatum, *L.*
Dry banks. Not rare.

Nabalus albus, *Hook.* (WHITE LETTUCE; RATTLESNAKE ROOT.)
Dry woods and banks. Frequent.

Var. serpentaria.
With the other form. Dr. Chas. Pickering considered the typical form as not found here; he had seen it in Canada.

Nabalus altissimus, *Hook.*
Woods and copses. Common.

Taraxacum Dens-leonis, *Desf.* (DANDELION.)
Everywhere that grass will grow. One of the earliest flowers, and prized by every child, often the first botanical inspiration. In flower from April to December (1879).

Lactuca Canadensis, *L.* (WILD LETTUCE.)
Lynnfield, Salisbury, West Newbury, Andover, etc. Common.

Var. intergrifolia, *Torrey & Gray.*
Swampscott (G. D. Phippen); Chebacco, etc. Not so abundant as the last form.

Var. sanguinea, *Torrey & Gray.*
Beverly, Boxford, Lynn (Tracy), etc. Not rare.

Mulgedium leucophæum, *DC.* (BLUE LETTUCE.)
Salem Mill pond borders, 1824 (Dr. Chas. Pickering); Beverly (G. D. Phippen); West Gloucester, Manchester, 1860 (C. H. Norris), etc. Frequent.

Sonchus oleraceus, *L.* (COMMON SOW-THISTLE.)
"Rare in Lynn" (Tracy). "A weed in my garden near the shore, Salem" (G. D. Phippen); Derby wharf, Salem, etc. Common. (Nat. from Eu.)

Sonchus asper, *Vill.*
"Rare in Lynn" (Tracy); Salem, Newbury, etc. Frequent. (Nat. from Eu.)

Sonchus arvensis, *L.*
Danvers (Oakes); Hamilton (G. D. Phippen). Quite common. (Nat. from Eu).

LOBELIACEÆ.

(LOBELIA FAMILY.)

Lobelia cardinalis, *L.* (CARDINAL-FLOWER.)
Frequent in wet places, particularly along brooks and borders of ponds. Several acres of this plant at Boxford, in the bed of a mill pond (1877). "Flowers red on a ground of white" (Proc. E. I. Vol. III, p. 110). "Pink, almost white in Haverhill" (Mrs. Downs).

Lobelia syphilitica, *L.* (GREAT LOBELIA.)
Reported at Andover, but as it is not mentioned otherwise, this may have been a mistake. It is probable that this species is known here only in cultivation (Salem, G. D. Phippen).

Lobelia inflata, *L.* (INDIAN TOBACCO.)
Dry places. Common.

Lobelia spicata, *Lam.*
Common in various soils, but generally in moist ones.

Lobelia Dortmanna, *L.* (WATER LOBELIA.)
Frequent in ponds, growing in about one foot of water. Martin's pond, Andover; Kenoza lake, Haverhill; Chebacco; Cape pond, Rockport. "Rare. Spring pond, Lynn" (Tracy), etc.

CAMPANULACEÆ.

(CAMPANULA FAMILY.)

Campanula rotundifolia, *L.* (HAREBELL.)
Banks, etc. Towns along the Merrimac valley.

Campanula aparinoides, *Pursh.*
Meadows, among other plants; in various parts of the county.

Campanula glomerata, *L.* (CLUSTERED BELLFLOWER.)
"Dark Lane, locality first noticed by Dr. Nichols" (Oakes in Hovey's Mag., Vol. VII); Topsfield, June, 1859 (Tracy, Proc. E. I., Vol. II, p. 390); "Danvers, Mass., etc." (Gray's Manual). The Dark Lane locality has probably been destroyed some time since. Danversport (J. H. Sears); Danvers, near the Asylum, 1879 (Walter Faxon). (Adv. from Eu.)

Campanula rapunculoides, *L.*
Frequent in grassy places and roadsides in the older towns. (Adv. from Eu.)

Specularia perfoliata, *A. DC.* (VENUS'S LOOKING-GLASS.)
Dry places. Frequent.

ERICACEÆ.

(HEATH FAMILY.)

Gaylussacia dumosa, *Torr. & Gray.*
Lynnfield (Dr. Chas. Pickering); Gloucester (Mrs. Kettle).

Gaylussacia frondosa, *Torr. & Gray.* (DANGLEBERRY.)
"Occasional in Lynn" (Tracy); Gloucester (Oakes); Magnolia Swamp, 1878; Middleton, Wenham Swamp, Amesbury, etc. Not very abundant.

Gaylussacia resinosa, *Torr. & Gray.* (BLACK HUCKLEBERRY.)
The market huckleberry.

Vaccinium Oxycoccus, *L.* (SMALL CRANBERRY.)
This has been reported in several places, yet no specimens have reached the P. A. S. herbarium. It should be looked for in peat bogs.

Vaccinium macrocarpon, *Ait.* (AMERICAN, OR LARGE CRANBERRY.
Very common in meadows. The finest locality noticed is a hollow among the dunes on Plum Island.

Vaccinium Vitis-Idæa, *L.* (COWBERRY.)
"First noticed by Oakes, in Danvers, about 1820" (Proc. E. I., Vol. I, p. 12). This locality is now exhausted, but Mr. J. H. Sears has found it growing more abundantly at another place. Bigelow (Fl. Bost. 2d ed., 1824), says, "In Lynn, Mr. Oakes;" which is copied into Hitchcock's Cat. An. & Pl., Mass. This is of course a mistake as Oakes' specimens are all from Danvers. It is corrected in the Fl. Bost. 3d ed., 1840. This "is the upland cranberry, whose berries are brought in quantities to the Boston market" (Dr. Chas. Pickering, Chron. Hist. Pl., p. 459).

Vaccinium Pennsylvanicum, *Lam.* (DWARF BLUEBERRY.)
The early blueberry.

Vaccinium vacillans, *Solander.* (LOW BLUEBERRY.)
Orne's Point, Salem (Dr. Chas. Pickering); Rockport, (C. W. Pool); Prospect Hill, Peabody; Newbury, Salisbury, etc. Frequent on wooded hills.

Vaccinium corymbosum, *L.* (SWAMP, OR HIGH-BUSH BLUEBERRY.)
The high blueberry varies exceedingly. The one commonly known by the name, being a tall shrub having large berries covered with a

bloom, the calyx remaining prominent upon the berry. Low forms with a similar berry are also found.

Var. atrococcum.
A very distinct form with smaller, flesh-colored flowers and very dark berries without a bloom. The calyx not prominent. Common.

Chiogenes hispidula, *Torr. & Gray.* (CREEPING SNOWBERRY.)
Pleasant pond, Wenham (Dr. Charles Pickering); Georgetown (Mrs. Horner); Essex and Middleton woods, etc. Rather rare.

Arctostaphylos Uva-ursi, *Spreng.* (BEARBERRY.)
Peabody, Gloucester, Andover, Lynn. Exposed rocky hills. Frequent. A variety with red flowers is collected by Mrs. Horner.

Epigæa repens, *L.* (MAY FLOWER.)
Near Ship-rock, Peabody, 1860; Lawrence; Gloucester (once very abundant); Topsfield, Boxford; several places in Salisbury (J. G. Whittier, Proc. E. I., Vol. III); Haverhill (Mrs. Downs); Marblehead Neck (Ed. Marblehead Messenger, G. F. Flint). The careless manner in which collectors gather this plant is the cause of its entire disappearance in many localities, and its scarcity in others.

Gaultheria procumbens, *L.* (CHECKERBERRY.)
Woods. Common.

Leucothoë racemosa, *Gray.*
"Near Mineral spring pond or by Tapley's brook" (Dr. Chas. Pickering); Danvers (Dr. Osgood's list); Magnolia Swamp, Gloucester (Proc. E. I., Vol. II, p. 35). Rare.

Cassandra calyculata, *Don.*
Borders of boggy ponds. Common. A charming spring flower, the buds opening in winter in the house, if collected in December.

Andromeda polifolia, *L.*
Cedar pond, Wenham, 1824 (Dr. Chas. Pickering). Rare. This station still exists.

Andromeda ligustrina, *Muhl.*
Common. Damp woods and roadsides.

Clethra alnifolia, *L.* (CLETHRA.)
Roadsides. Common. A form only "a few inches high with racemes of full size" was found at Rockport by Mr. G. D. Phippen.

Calluna vulgaris, *Salisb.* (HEATHER.)
Discovered at Tewksbury (see Am. Nat. Vol. X, Aug. 1876) a very short distance from the county line by Jackson Dawson, about 1865. This station would be considered small even for the most scarce plants. In 1876, a single plant was noticed, in Andover, by William

Mitchell. Through the kindness of Professor Goldsmith this locality was visited June, 1879. The place where this specimen grows has been ploughed or overturned certainly within twenty years. Plants from the Tewksbury locality have been successfully cultivated in Andover for some years, doing well in ordinary gardens. It is hard to imagine this a truly native plant from the Essex county stations, but the species is found at "Cape Elizabeth, Maine, and less rare in Nova Scotia, Cape Breton and Newfoundland" (Gray Fl. N. A.).

Kalmia latifolia, *L.* (MOUNTAIN LAUREL.)
Most of the towns in the county. More abundant on Cape Ann and in the Merrimac valley. A plant having double flowers was found, at Gloucester, by Mr. G. D. Phippen. In clearings, the exposed plants have very pink flowers.

Kalmia angustifolia, *L.* (LAMBKILL.)
Fields and pasture lands. Common.

Kalmia glauca, *Ait.* (PALE LAUREL.)
Gloucester (Dr. Chas. Pickering); Cedar pond, Wenham (S. P. Fowler, Proc. E. I., Vol. I). Not since found, and not represented by county specimens in the P. A. S. herbarium. The Cedar pond station is one that this plant would usually select.

Rhododendron viscosum, *Torrey.* Azalea viscosa, L. (Gray's Manual.) (SWAMP HONEYSUCKLE.)
Wet places. Common.

Var. **glauca.**
Ipswich (Oakes); borders of Wenham pond. Frequent.

Rhododendron nudiflorum, *Torr.*, Azalea nudiflora, *L.* (Gray's Manual.) (PURPLE AZALEA.)
"Banks of the Merrimac; West Parish, Andover" (Mrs. S. M. Downs). As this species is abundant in New Hampshire, it ought to be quite frequently met with north of the Merrimac.

Rhododendron Rhodora, *Don.* Rhodora Canadensis, *L.* (Gray's Manual). (RHODORA.)
Salem Great Pastures, locality thought now to be exhausted (David Waters); Ipswich (Oakes); Salisbury (J. G. Whittier, Proc. E. I., Vol. III); Peabody, Haverhill, Andover, Boxford (some plants five and six feet high). "One place in Lynn" (Tracy), etc. Frequent.

Ledum latifolium, *Ait.* (LABRADOR TEA.)
"Essex County, 1824" (memo. Dr. Chas. Pickering). Not in the P. A. S. collection from the county.

Pyrola rotundifolia, *L.*
Rather deep woods. Common.

Pyrola elliptica, *Nutt.* (SHIN-LEAF.)
Woods. Common.

Pyrola chlorantha, *Swartz.*
"Rare in Lynn" (Tracy). Common at Beverly, Essex, etc.

Pyrola secunda, *L.*
"Rare in Lynn" (Tracy). Most abundant at Middleton, Georgetown, Essex, Haverhill.

Moneses uniflora, *Gray.*
"Riall side" Beverly, 1850 (Gilbert Streeter and H. J. Cross). "Mr. Oakes has sent it from Lynn" (Bigelow's Fl. Bost., 2d ed., 1824). "Mr. Oakes has sent it from Wenham" (Bigelow's Fl. Bost., 3d ed., 1840). The last is evidently a correction of the former statement. Salisbury and Amesbury (G. Markoe and S. Shepard). "Rare, only from Saugus" (Tracy). The richest locality for this plant thus far observed is in the Crooked pond woods, Boxford, where the specimens are very fine and abundant.

Chimaphila umbellata, *Nutt.* (PIPSISSEWA; PRINCE'S PINE.)
Woods. Common.

Chimaphila maculata, *Pursh.* (SPOTTED WINTERGREEN.)
Pleasant pond woods, Wenham (J. R.); Middleton (Tracy); Beverly (Mr. J. Redmond); Andover (Mrs. Downs). Rare.

Monotropa uniflora, *L.* (INDIAN PIPE.)
Woods. Common.

Monotropa Hypopitys, L. (PINE-SAP.)
Frequent in situations similar to the last. Mrs. Horner finds in Georgetown a variety with red flower.

ILICINEÆ.

Aquifoliaceæ (Gray's Manual).

(HOLLY FAMILY.)

Ilex opaca, *Ait.* (AMERICAN HOLLY.)
"Dogtown Commons, Rockport" (Frank Lufkin). The trees are of considerable size. The only locality thus far reported; its origin only conjectural.

Ilex verticillata, *Gray.* (BLACK ALDER.)
Damp roadsides by walls. Common.

Ilex lævigata, *Gray.* (SMOOTH WINTERBERRY.)
Not nearly so abundant as the last. Cedar pond, Lynn (H. A. Young); Middleton, Georgetown, etc.

Ilex glabra, *Gray.* (INKBERRY.)
Magnolia Swamp, Wenham Swamp, Rockport (C. W. Pool), etc. Not very common, yet abundant where it grows at all.

Nemopanthes Canadensis, *DC.* (MOUNTAIN HOLLY.)
Wet places in the woods. Quite common.

PLANTAGINACEÆ.

(PLANTAIN FAMILY.)

Plantago major, *L.* (COMMON PLANTAIN.)
A cosmopolitan weed.

Plantago Rugelii, *Decaisne.* (RED STEMMED PLANTAIN.)
This species, which nearly every one takes to be a variety of the common plantain, is not found in the old world, but is very common here, and is considered to be a native plant. The leaves are light green and smooth, the capsules longer than those of the last, and the bases of the petioles are distinctly red. It will be found in Gray's Manual under P. Kamtschatica.

Plantago decipiens, *Barneoud.* Plantago maritima, *L.*, var. juncoides. (Gray's Manual.) (SEASHORE PLANTAIN.)
Frequent on the shore on rocks, sand and salt marshes. The true P. maritima grows north of the St. Lawrence and on the Pacific coast.

Plantago lanceolata, *L.* (RIBGRASS; ENGLISH PLANTAIN.)
Common in fields. (Nat. from Eu.)

PLUMBAGINACEÆ.

(LEADWORT FAMILY.)

Statice Limonium, *L.*, var. **Caroliniana.** (MARSH-ROSEMARY.)
Salt marshes. Common. Much used in a dry state for winter decorations. It is now sold at the street corners in Boston with the Lygodium and grasses.

PRIMULACEÆ.

(PRIMROSE FAMILY.)

Trientalis Americana, *Pursh.* (STAR-FLOWER.)
Woods. Common.

Lysimachia thyrsiflora, *L.* (TUFTED LOOSESTRIFE.)
Borders of ponds. Not very common.

Lysimachia stricta, *Ait.*
Wet places. Common. A variety, bearing numerous bulblets in the axils of the leaves, is found in more shady places.

Lysimachia quadrifolia, *L.*
Banks and copses. Common.

Lysimachia vulgaris, *L.*
Roadside in Beverly. A garden plant escaped. (Adv. from Eu.)

Lysimachia nummularia, *L.* (MONEYWORT.)
Cultivated in old gardens, occasionally escaping. Peabody, Salem, etc. (Adv. from Eu.)

Steironema ciliatum, *Raf.*, Lysimachia ciliata, *L.* (Gray's Manual.)
Chebacco; towns in the Merrimac valley, etc.; abundant. Rare in the southern and eastern portions of the county.

Steironema lanceolatum, *Gray*, Lysimachia lanceolata, *Walt.* (Gray's Manual.)
North Andover, Haverhill, etc.

Var. hybridum.
Lynnfield, Brookdale, Lynn, Middleton, etc.

Glaux maritima, *L.* (SEA-MILKWORT.)
Ipswich (Oakes); "in wet salt marshes about Salem" (Dr. Chas. Pickering).

Anagallis arvensis, *L.* (PIMPERNEL; POOR MAN'S WEATHER-GLASS.)
Fields near the shore. Common. (Nat. from Eu.)

Hottonia inflata, *Ell.* (FEATHERFOIL.)
Beverly; "frequent in Breed's pond, Lynn" (Tracy); Lanesville (Mrs. Babson); Andover (Mrs. Downs); Boxford, etc. In slow streams and pond holes. Not very common.

LENTIBULACEÆ.

(BLADDERWORT FAMILY.)

Utricularia inflata, *Walt.*
Peabody, in a pond on the road from Salem to Lynnfield Hotel; Lynnfield (G. D. Phippen); Andover (Rev. H. P. Nichols); Lily pond, Lynn (Tracy); Boxford, etc. Rather rare.

Utricularia vulgaris, *L.*, var. **Americana.** (COMMON BLADDERWORT.)
Ponds and ditches. Common. The plants float about the larger ponds, collecting in quiet nooks to flower.

Utricularia intermedia, *Hayne.*
Topsfield; "Bowler swamp, Lynn" (Tracy); Andover (Mrs. Downs); Chebacco, Haverhill, Lawrence, Methuen, etc. Not rare.

Utricularia gibba, *L.*
Chebacco pond, Aug. 1876 (J. R.); Georgetown (Mrs. Horner). Rare.

Utricularia purpurea, *Walt.*
"Hawks' pond, Lynnfield, July, 1858" (G. D. Phippen, Proc. E. I. Vol. II, p. 293). Hamilton (G. D. Phippen); Lynnfield Hotel, little pond; Haverhill (Mrs. Downs); Chebacco (not in flower); "Danvers, Dr. Nichols" (Bigelow's Fl. Bost. 2d ed., 1824). Rare.

Utricularia resupinata, *Greene.*
"Discovered at Tewksbury by B. D. Greene" (Bigelow's Fl. Bost., 3d ed., 1840); Haggett's pond, Andover, 1875 (C. E. Faxon, Geo. E. Davenport) and same locality reported by Mr. Wm. Boott. It seems probable that Bigelow's reference should be to the Haggett's pond locality as it is quite near Tewksbury, and that Mr. Boott's information came from Mr. Greene. This species is also collected at Boxford (Mrs. Wilmarth).

Utricularia cornuta, *Michx.*
Borders of ponds in the sand. Frequent.

BIGNONIACEÆ.

(BIGNONIA FAMILY.)

Tecoma radicans, *Juss.* (TRUMPET FLOWER.)
A native westward, extensively cultivated, not always proving hardy in exposed situations.

Catalpa bignonioides, *Walt.* (COMMON CATALPA.)
A native of the south, early introduced here as a shade and ornamental tree, it is now likely to be replaced by the western Catalpa (C. speciosa), which closely resembles it, but is much more hardy. The wood is extremely durable.

Martynia proboscidea, *Glox.*
A native westward, much cultivated here, the fruit being used for pickling.

OROBANCHACEÆ.

(BROOM-RAPE FAMILY.)

Epiphegus Virginiana, *Bart.* (BEECH-DROPS.)
Frequent under beeches. The Chebacco region is a fine locality.

Aphyllon uniflora, *Torr. & Gray.*
"North Salem, 1827" (memo. Rev. J. L. Russell); Derby estate, Salem, 1875 (Hugh Wilson); Saugus (J. H. Emerton); Andover, Georgetown, Amesbury, Merrimac; Boxford (Miss M. E. Perley), Middleton, etc. Frequent in thin woods.

SCROPHULARIACEÆ.

(Figwort Family.)

Verbascum Thapsus, *L.* (Common Mullein.)
Roadsides and fields. Common. (Nat. from Eu.)
Verbascum Blattaria, *L.* (Moth Mullein.)
"By European colonists, was carried prior to 1670 (Josselyn) to northeast America" (Dr. Chas. Pickering, Chron. Hist. Pl., p. 464), and yet it never seems to have been very common. "By the roadsides in Lynn" (Dr. Cutler, 1783, in Mem. Am. Acad., Vol. I); West Newbury (Wm. Merrill); Wenham (W. P. Conant); "Danvers in fields" (G. D. Phippen, Dr. Osgood's list, 1853). (Nat. from Eu.)
Verbascum nigrum, *L.*
Tapleyville (J. H. Sears). Probably introduced with wool. (Adv. from Eu.)

Linaria Canadensis, *Spreng.*
Cart-paths. Common.
Linaria vulgaris, *Mill.* (Toad-Flax; Butter-and-eggs.)
Fields and roadsides. Frequent. (Nat. from Eu.)
Linaria Elatine, *Mill.*
"Naturalized at Ipswich" (Oakes in Hovey's Mag., Vol. XIII). (Adv. from Eu.)

Antirrhinum majus, *L.* (Large Snapdragon.)
Once established in a garden, it is sure to remain; it also occasionally escapes into the roadsides near old country gardens. (Adv. from Eu.)

Scrophularia nodosa, *L.* (Figwort.)
"Boston street, between Washington street and the paper factory; Lynn" (Tracy); Beverly Farms (G. D. Phippen).

Chelone glabra, *L.* (Turtle-head; Snake-head.)
Near brooks. Frequent.

Mimulus ringens, *L.* (Monkey-Flower.)
Damp ground. Common.

Gratiola Virginiana, *L.*
Lynn, near Flax pond (Tracy, H. A. Young); Lawrence (J. R.); Andover (Mrs. Downs). Scarce.

Gratiola aurea, *Muhl.*
Shores of ponds and swamps. Common. There is a white variety of this species which grows in Bowler Swamp, Lynn.

Ilysanthes gratioloides, *Benth.* (FALSE PIMPERNEL.)
Bowler Swamp, Lynn; Chebacco, North Andover, Gloucester, etc. Frequent.

Veronica Anagallis, *L.* (WATER SPEEDWELL.)
"So thick in many ditches as almost to choke them up. Margins of Breed's pond and similar boggy localities" (Tracy's Lynn Flora).

Veronica Americana, *Schweinitz* (V. Beccabunga of Am. Authors).
"Outlet of Mineral spring pond. Perhaps the locality referred to by Higginson" (memo. Dr. Chas. Pickering).

Veronica scutellata, *L.* (MARSH SPEEDWELL.)
Among other plants in wet places.

Veronica serpyllifolia, *L.*
Low roadsides and other moist places.

Veronica peregrina, *L.*
A weed in damp places.

Veronica arvensis, *L.* (CORN SPEEDWELL.)
"Rare. Holmes's mill, Walnut street, Lynn" (Tracy); "Salem Great Pastures, 1842" (memo. Rev. J. L. Russell); Danvers (Dr. Osgood's list); Essex County (Oakes). Not common. (Nat. from Eu.)

Gerardia purpurea, *L.*, var. **paupercula** (Gray's Fl. N. A.).
Common in wet pastures. Rarely over eight inches high. A variety with white flowers was collected in Salem by Miss S. Russell.
"G. purpurea of eastern Massachusetts is probably distinct from that bearing the same name in our Middle States; observed around Philadelphia two feet high with flowers twice as large" (Dr. Chas. Pickering in a letter of July 15, 1875). The typical form does not seem to have been observed here thus far by collectors.

Gerardia maritima, *Raf.*
Salem, Salisbury, West Gloucester, etc. Salt marshes. Not rare.

Gerardia tenuifolia, *Vahl.*
Dry hillsides. Frequent.

Gerardia flava, *L.*, partly.
Lynn (Tracy), Rockport (Pool), Beverly, Andover, Amesbury, Newbury, etc. Frequent on wooded hillsides.

Gerardia quercifolia, *Pursh.*
Danvers, Ipswich (Oakes), Andover, Methuen, Haverhill, etc. In similar places to the last.

Gerardia pedicularia, *L.*
Dry thin woods, Hamilton, Lynn, Danvers, Linebrook (Miss Perley), Middleton, Topsfield, etc. As the last two, frequent.

Castilleia coccinea, *Spren.* (PAINTED-CUP.)
Danvers, Boxford, Topsfield, Andover, North Andover, and towns in that region. Rare in other parts of the county. A variety with more finely cut leaves and yellowish flowers is found in Danvers by J. H. Sears.

Rhinanthus Crista-galli, *L.* (COMMON YELLOW-RATTLE.)
"Near Salem" (probably about 1824) and elsewhere (Dr. Chas. Pickering Chron. Hist. Pl. p. 501); Gloucester (Mrs. Babson, and herb., P. A. S.); "Haverhill" (Mrs. Downs); Georgetown (Mrs. C. N. S. Horner). As considered in the Plymouth locality, it is probably introduced here from farther north.

Pedicularis Canadensis, *L.* (LOUSEWORT.)
Copses, etc. Frequent. "It will not live in cultivation when transplanted, perhaps a parasite" (memo. G. D. Phippen).

Melampyrum Americanum, *Michx.* (COW-WHEAT.)
Dry woods. Common.

VERBENACEÆ.

(VERVAIN FAMILY.)

Verbena hastata, *L.* (BLUE VERVAIN.)
"Very rare in Lynn" (Tracy). Most abundant in the central and northwestern portions of the county where it is common.

Verbena urticifolia, *L.* (WHITE VERVAIN.)
Roadsides. Frequent.

Phryma Leptostachya, *L.* (LOPSEED.)
Byfield (Rev. A. P. Chute), Groveland, Georgetown (Mrs. Horner), Danvers. Not very common.

LABIATÆ.

(MINT FAMILY.)

Teucrium Canadense, *L.* (WOOD SAGE.)
Saugus river banks, Gloucester, North Andover, Byfield, Newbury, etc. Not very common.

Trichostema dichotomum, *L.* (BLUE CURLS.)
Dry sterile fields. Common.

Mentha viridis, *L.* (SPEARMINT.)
Danvers (Dr. Osgood's list), Rockport (Pool), West Gloucester, Groveland, etc. Frequent. (Nat. from Eu.)

Mentha sativa, *L.* (WHORLED MINT.)
Chebacco road, Hamilton, Aug. 24, 1874. (Adv. from Eu.)

Mentha piperita, *L.* (PEPPERMINT.)
Rockport (Pool), Nahant, Danvers, Wenham, etc. Occasional. (Nat. from Eu.)

Mentha Canadensis, *L.* (WILD MINT.)
Meadows. Common.

Lycopus Virginicus, *L.* (BUGLE-WEED.)
"Near Salem" (S. B. Buttrick, Proc. E. I. Vol. II, p. 241), Beverly (John C. Phillips), Essex County (Dr. Chas. Pickering). Seemingly not very common.

Lycopus sinuatus, *Ell.*, L. Europæus, var. sinuatus (Gray's Manual). Moist ground. Common and quite variable. A native species. Lycopus Europæus, *L.*, appears only in this county as an adventitious plant and probably does not grow here.

Hyssopus officinalis, *L.*
Escaped from old gardens in a few places. Georgetown (Mrs. Horner). (Adv. from Eu.)

Pycnanthemum muticum, *Pers.* (MOUNTAIN MINT.)
"Very rare, stone barn, Swampscott" (Tracy); Danvers (J. H. Sears; and a few other places in the county. Scarce.

Pycnanthemum lanceolatum, *Pursh.*
Georgetown (Mrs. Horner), and the towns in that neighborhood. Scarce.

Origanum vulgare, *L.* (WILD MARJORAM.)
Georgetown (Oakes), Rowley (Rev. J. L. Russell). Not common. (Nat. from Eu.)

Thymus Serpyllum, *L.* (CREEPING THYME.)
"Dry pastures, Boxford, 1819, Oakes and Osgood" (Oakes in Hovey's Mag. Vol. VII). (Adv. from Eu.)

Thymus vulgaris, *L.* (GARDEN THYME.)
Cultivated in gardens, occasionally escaping. Beverly (W. D. Silsbee).

Hedeoma pulegioides, *Pers.* (AMERICAN PENNYROYAL.)
"Clearly the 'peneriall' seen by W. Wood l. 5—and Higginson, eastern Massachusetts, and 'upright peniroyal' of Josselyn Rar. 44."

(Dr. Chas. Pickering, Chron. Hist. Pl. p. 944). Frequent in old fields and rocky places. A native plant.

Collinsonia Canadensis, *L.* (RICH-WEED; STONE-ROOT.)
Ipswich (Oakes), Georgetown (Mrs. Horner). Scarce.

Monarda didyma, *L.* (OSWEGO TEA.)
Chebacco road, Georgetown, Haverhill, and often in cultivation. Not common in a wild state.

Monarda fistulosa, *L.* (WILD BERGAMOT.)
Georgetown (Mrs. Horner), "Danvers" (Dr. Osgood's list), North Andover (Mrs. A. J. Haskell). Not very common and hardly a native of this region, but probably introduced from farther north.

Nepeta Cataria, *L.* (CATNIP.)
Common. (Adv. from Eu.)

Nepeta Glechoma, *Benth.* (GROUND IVY.)
Damp places. Frequent. (Adv. from Eu.)

Physostegia Virginiana, *Benth.*
Roadside at Middleton, Aug. 1874 (J. R.); Danvers, 1853 (Dr. Osgood's list); Hamilton (G. D. Phippen); Topsfield. Introduced from farther west (N. Y., etc.) for cultivation, whence it has escaped.

Brunella vulgaris, *L.* (SELF-HEAL.)
Fields, etc. Common. A few plants having white flowers were found at North Reading, 1874 (J. R.).

Scutellaria galericulata, *L.* (SKULLCAP.)
Damp places. Frequent. A quite rigid form grows near the seashore, noticed at "Pebblestone beach," Rockport, 1875 (J. R.).

Scutellaria lateriflora, *L.* (MAD-DOG SKULLCAP.)
Wet places. Most common in the interior of the county.

Marrubium vulgare, *L.* (COMMON HOREHOUND.)
Danvers (Oakes). (Nat. from Eu.)

Galeopsis Tetrahit, *L.* (COMMON HEMP-NETTLE.)
Ipswich (Oakes), Gloucester (Mrs. Babson), Wenham, shores of the Merrimac at West Newbury, Boxford, etc. Rather scarce. (Nat. from Eu.)

Galeopsis Ladanum, *L.* (RED HEMP-NETTLE.)
"Chelsea beach by Dr. Bigelow" (Dr. Chas. Pickering); and since at the same place (H. A. Young). Although not in the geographical limits of the county it comes properly into the flora of the region as it is within our latitude, and but a few rods beyond the county line.

Stachys palustris, *L.* (HEDGE NETTLE.)
"Naturalized at Ipswich, exactly like the European plant" (Oakes in Hovey's Mag., Vol. XIII).

Stachys aspera, *Michx.*, S. palustris, var. aspera (Gray's Manual).
"Sparingly at Lynn" (Tracy).

Stachys cordata, *Riddell*, S. palustris, var. cordata (Gray's Manual).
Specimens collected at Danvers, by Mr. Sears, were referred here by Prof. Watson.

Stachys hyssopifolia, *Michx.*
"Appeared mysteriously on the wall at Paradise, Salem, 1824" (Dr. Chas. Pickering).

Leonurus Cardiaca, *L.* (MOTHERWORT.)
Country yards, etc. Common. (Nat. from Eu.)

Lamium amplexicaule, *L.*
"A weed in some old gardens in Lynn" (Tracy), and old gardens elsewhere. (Adv. from Eu.)

Lamium album, *L.* (WHITE DEAD-NETTLE.)
Old gardens, and now and then escaped. "L. rugosum (L. album, L.) grows freely on the cinder banks at Saugus" (Tracy, Proc. E. I., Vol. II, p. 389). (Adv. from Eu.)

Ballota nigra, *L.* (BLACK HOREHOUND.)
"Dark lane," Salem (memo. Rev. J. L. Russell). Rare. Locality now destroyed. (Adv. from Eu.)

BORRAGINACEÆ.

(BORAGE FAMILY.)

Asperugo procumbens, *L.*
A foreign plant, found on the road-bed of the E. R. R. near Salem, by Rev. J. L. Russell, some twenty years since.

Echium vulgare, *L.* (BLUE-WEED.)
Danvers; Lynn, 1838 (memo. Rev. J. L. Russell); Middleton, near the paper mill (G. D. Phippen); Salem (Mrs. Kimball); Beverly, near the old rubber factory, etc. Not very common. (Nat. from Eu.)

Lycopsis arvensis, *L.* (SMALL BUGLOSS.)
Ipswich (Oakes), "Andover" (Mrs. Downs). (Adv. from Eu.)

Symphytum officinale, *L.* (COMMON COMFREY.)
The specimen in the P. A. S. county herbarium is from Wm. P. Richardson, a botanist, who collected the plants of this region some

forty years ago. The precise locality is unknown. Noticed in Andover, June, 1879 (J. R.); Boxford (Miss Perley). (Adv. from Eu.)

Myosotis laxa, *Lehm.*, Myosotis palustris, var. laxa. (Gray's Manual.) (FORGET-ME-NOT.)
Salem Great Pastures; Lynn (Tracy); Ipswich, Hamilton, etc. Not rare in wet places.

Myosotis arvensis, *Hoffm.*
Noticed as a weed in a garden in Salem in 1874. It very probably was introduced with the packing around plants. No specimens were preserved.

Myosotis verna, *Nutt.*
Dry hills. Common.

Echinospermum Lappula, *Lehm.*
Salem (Mrs. Crosby), Essex County (Dr. Chas. Pickering). (Nat. from Eu.)

Echinospermum Virginicum, *Lehm.*, Cynoglossum Morisoni, *DC.* (Gray's Manual.) (BEGGAR'S LICE.)
North Andover (Rev. J. L. Russell), Boxford, Groveland, West Newbury, Amesbury, etc. Along walls. Quite a common plant.

Cynoglossum officinale, *L.* (HOUND'S TONGUE.)
"Old mill-site, Middleton, June, 1859" (Tracy, Proc. E. I. Vol. II, p. 383).

POLEMONIACEÆ.

(PHLOX FAMILY.)

Phlox paniculata, *L.* (GARDEN PHLOX.)
A native farther westward. Escaped in Beverly and two or three other places along the roadsides.

CONVOLVULACEÆ.

(CONVOLVULUS FAMILY.)

Ipomœa purpurea, *Lam.* (MORNING GLORY.)
Common in yards and gardens where it propagates itself by seeds. Derby Wharf, Salem; roadside in Wenham and other places. Escaped. (Adv. from Trop. Amer.)

Convolvulus arvensis, *L.* (BINDWEED.)
A pretty weed quite abundant from Rockport to Andover. (Nat. from Eu.)

Convolvulus sepium, *L.*, Calystegia sepium, *R. Brown.* (Gray's Manual.) (HEDGE BINDWEED.)
Commonly climbing over bushes in damp places. Often cultivated. Quite variable in the color of the flowers.

Convolvulus spithamæus, *L.*, Calystegia spithamæa, *Pursh.* (Gray's Manual.)
Reported from two localities but not represented in the P. A. S. collection. The identification is a little uncertain.

Cuscuta Epilinum, *Weihe.* (FLAX DODDER.)
"Old flax fields, Rowley, 1826;" (Oakes in Hovey's Mag. Vol. XIII). "Well known to farmers who have their flax fields greatly injured by it" (Dr. Cutler, 1783, in Mem. Am. Acad., Vol. I). (Adv. from Eu) Not now observed, flax being seldom cultivated.

Cuscuta Gronovii, *Willd.*
Common on herbaceous plants in damp places. "Common in hedges, etc." (Dr. Cutler, 1783, in Mem. Am. Acad., Vol. I).

SOLANACEÆ.

(NIGHTSHADE FAMILY.)

Solanum Dulcamara, *L.* (BITTERSWEET.)
"By European colonists, was carried to northeast America (later perhaps than 1670, for it is not mentioned by Josselyn), and has become completely naturalized." (Dr. Chas. Pickering, Chron. Hist. Pl. p. 531).
Common in damp places. Often called "Deadly Nightshade" in this region, which name belongs more properly to the next. Neither species is poisonous to handle as is often imagined.

Solanum nigrum, *L.* (COMMON NIGHTSHADE.)
Gloucester, Andover (Mrs. Downs), Salem (in two places); Newburyport, etc. Quite common. The American form of this species was thought by Dr. Pickering, from specimens collected about 1824, to be considerably different from the European plant, and Gray's new Flora of N. A. gives this as a native species.

Solanum rostratum, *Dunal.*
Rockport (Mrs. Wheeler). An immigrant from the southwest.

Physalis Alkekengi, *L.* (STRAWBERRY TOMATO; GROUND CHERRY.)
Frequently cultivated and occasionally spontaneous. "Salem Neck, near a compost heap" (memo. G. D. Phippen). Mr. Phippen also states that in Salem, some years since, a person styling himself the Rev.———sold the fruit of this plant at fifty cents a berry attributing to it some rare virtue and actually found customers. (Adv. from Eu.)

Physalis Virginiana, *Mill.* (Gray's Fl. N. A.), P. viscosa, *L.* (Gray's Manual.)
Near the cemetery, Lynn (C. E. Faxon). Probably introduced from farther west.

Nicandra physaloides, *Gærtn.* (APPLE OF PERU.)
"Rare, yards in Summer St., Lynn" (Tracy); Rockport (Pool); West Newbury (Wm. Merrill); Boxford (Miss Perley); Andover (Mrs. Downs). Scarce. (Adv. from Peru.)

Lycium vulgare, *Dunal.* (MATRIMONY-VINE.)
Escaped by a wall in Beverly, and a few other places. Common in old gardens. (Adv. from Eu.)

Hyoscyamus niger, *L.* (BLACK HENBANE.)
Apparently common in 1783 (Dr. Cutler in Proc. Amer. Acad., Vol. I). "Short beach, Nahant, often disappearing for a while" (Tracy, Proc. E. I. Vol. II, p. 275); Essex County (Oakes); "Ipswich near the depot, and the south side of Phillip's wharf, Salem" (memo. G. D. Phippen). (Adv. from Eu.)

Datura Stramonium, *L.* (COMMON STRAMONIUM.)
Common in waste places. (Adv. from Asia.)

Datura Tatula, *L.* (PURPLE THORN-APPLE.)
Often with the other (probably adv. from Trop. Amer.). Both are most frequent near the seashore.

Nicotiana rustica, *L.* (WILD TOBACCO.)
"Near Danvers" (Dr. Osgood's list). Not represented in the P. A. S. herbarium of Essex County plants, and possibly extinct, though once frequent according to early authors. (Adv. from Trop. Amer.)

Petunia nyctaginiflora.
On a waste heap, Nahant, Aug., 1878; Wenham, by the roadside, etc. Not permanently established, although very common in cultivation. (Adv. from So. America.)

GENTIANACEÆ.

(GENTIAN FAMILY.)

Gentiana crinita, *Frœl.* (FRINGED GENTIAN.)
Beverly, scarce; Hamilton, scarce; more abundant in the northern and western portions of the county. Mrs. Horner collected in Georgetown a single plant hardly less than fifteen inches high, having thirty flowers. Rev. H. P. Nichols states that the variety with white flowers and pale green leaves is quite constant in the Andover region.

Gentiana Andrewsii, *Griseb.* (CLOSED GENTIAN.)
Quite often found in the region of the last, yet nowhere very common. "The soapwort species (G. saponaria)" referred to in the Proc. E. I., Vol. II, p. 406, as growing in Groveland, is undoubtedly a mistake, and should be this species of which G. saponaria, Frœl, is a synonyme.

Bartonia tenella, *Muhl.*
Lynnfield (Rev. A. P. Chute, Proc. E. I., Vol. II, p. 47). "Swamp at the Pickman Farm, Salem" (memo. Dr. Chas. Pickering).

Menyanthes trifoliata, *L.* (BUCKBEAN.)
Bogs in the central and northern parts of the county, where it is common. Scarce in most other places. Bog-bean would be a better name.

Limnanthemum lacunosum, *Griseb.* (FLOATING HEART.) Quite common in ponds.

APOCYNACEÆ.

(DOGBANE FAMILY.)

Apocynum androsæmifolium, *L.* (SPREADING DOGBANE.)
Frequent. Roadsides and copses. A variety with downy leaves and quite deep rose-colored flowers is found at Georgetown by Mrs. Horner.

Apocynum cannabinum, *L.* (INDIAN HEMP.)
Georgetown (Mrs. Horner); Sluice pond, Lynn (Herbert A. Young); Andover (Mrs. Downs). Shores of the Merrimac at W. Newbury.

Var. **hypericifolium** (Gray's Manual).
Essex County (Oakes). "Oakes showed me a depressed form at Middleton pond" (Dr. Chas. Pickering). This species is much rarer than the first.

ASCLEPIADACEÆ.

(MILKWEED FAMILY.)

Asclepias Cornuti, *Decaisne.* (COMMON MILKWEED OR SILKWEED.)
Very common. Varies much, both as regards the shape of the leaves and the color of the flowers.

During the last century the coma of the seeds of this plant was used for wick-yarn. "The candles will burn equally free and afford a clearer light than those made of cotton wicks. They will not require so frequent snuffing and the smoke of the snuff is less offensive" (Dr. Cutler, 1783, in Proc. Am. Acad., Vol. I).

About 1838 a patent was granted Miss Margaret Gerrish, of Salem, for a process by which the fibre of this species of milkweed was used for the manufacture of various kinds of thread, cloth, etc. The specimens illustrating this are now at the Essex Institute.

Asclepias phytolaccoides, *Pursh.* (POKE MILKWEED.)
Frequent in the Merrimac valley. Scarce in the southern towns.

Asclepias purpurascens, *L.*
"Danvers, July 17, 1853" (Dr. Osgood's list). "North Andover, in the rear of the Osgood estate" (Dr. Chas. Pickering), Middleton, 1880 (J. R.). Rare.

Asclepias quadrifolia, *Jacq.*
"Dungeon pasture, Lynn, rare" (Tracy). "Kernwood, Salem" (G. D. Phippen). Danvers (J. H. Sears), Middleton, etc. Scarce.

Asclepias incarnata, *L.*, var. **pulchra,** *Pers.* (SWAMP MILKWEED.)
Wet places. Frequent. Cultivated by Mr. G. D. Phippen in Salem, who states that "the plants grew freely from seeds, the offspring being smoother."

Asclepias obtusifolia, *Michx.*
Lynnfield (Rev. A. P. Chute, Proc. E. I., Vol. II, p. 37), "Rockport" (Pool); Ipswich (Oakes). Rare.

Asclepias tuberosa, *L.* (BUTTERFLY-WEED. PLEURISY-ROOT.)
Wenham; Danvers; Topsfield; and the Merrimac valley towns, frequent. Rare elsewhere. All the native milkweeds are interesting in cultivation. Mr. G. D. Phippen of Salem has experimented with several, the most satisfactory being A. tuberosa, one plant of which bore, in his garden, one hundred clusters of flowers.

OLEACEÆ.

(OLIVE FAMILY.)

Ligustrum vulgare, *L.* (PRIVET; PRIM.)
Rial Side, Beverly; Lynn, Wenham, Andover, etc. Common. "Known in a wild state as long ago as when Dr. Manasseh Cutler botanized here in 1790" (memo. G. D. Phippen). Adv. from Eu., but naturalized in various parts of the county.

Syringa vulgaris, *L.* (LILAC.)
Cultivated in every old garden, often found escaped, although not far from dwellings. The white variety is less frequent than the ordinary form. (Adv. from Eu.)

Fraxinus Americana, *L.* (WHITE ASH.)
Common in most parts of the county, but seldom in groves.
Fraxinus pubescens, *Lam.* (RED ASH.)
Frequent, riverbanks, etc., in the central portion of the county.
Fraxinus sambucifolia, *Lam.* (BLACK ASH; WATER ASH.)
Swampy woods. Frequent.

ARISTOLOCHIACEÆ.

(BIRTHWORT FAMILY.)

Asarum Canadense, *L.* (WILD GINGER.)
This grows in Andover and Haverhill, and has been cultivated by Mr. S. P. Fowler in Danvers. Introduced from beyond the county limits.

PHYTOLACCACEÆ.

(POKEWEED FAMILY.)

Phytolacca decandra, *L.* (GARGET; POKE.)
Damp ground and freshly cleared land. Common. The root is poisonous, yet "the stalks are in early spring sometimes eaten as a substitute for asparagus." (Gray's Manual.) For cultivation, where it can have sufficient room, this makes a most striking autumn plant.

CHENOPODIACEÆ.

(GOOSEFOOT FAMILY.)

Chenopodium album, *L.* (PIGWEED.)
A weed everywhere. (Nat. from Eu.)
Chenopodium glaucum, *L.* (OAK-LEAVED GOOSEFOOT.)
"Marblehead Neck, Sept. 8, 1858" (Tracy, Proc. E. I., Vol. II, p. 312); Salem, 1879; Lynn (Herbert A. Young). (Nat. from Eu.)
Chenopodium urbicum, *L.*
Near Flax pond, Lynn, 1879 (H. A. Young); "occasional in Lynn" (Tracy). This species is often confused with the next, forms of which are taken for it. (Nat. from Eu.)
Chenopodium hybridum, *L.* (MAPLE-LEAVED GOOSEFOOT.)
Frequent. Both the small form and the large are with an extensive spreading panicle of flowers. (Nat. from Eu.)
Chenopodium Botrys, *L.* (JERUSALEM OAK; AMBROSIA.)
A garden plant or weed, often seen in yards; Lynn, 1879 (H. A. Young), etc. (Adv. from Eu.)

Chenopodium ambrosioides, *L.*, var. **anthelminticum,** *Gray.* (WORMSEED.)
Common in Lynn where it seems thoroughly established. (Nat. from Trop. Am.)

Blitum maritimum, *Nutt.* (COAST BLITE.)
"Near the mouth of Saugus river" (Dr. Chas. Pickering). "Annisquam" (Mrs. Downs).

Blitum capitatum, *L.* (STRAWBERRY BLITE.)
A native of the west, cultivated and perpetuating itself by seeds in gardens and yards.

Atriplex patula, *L.*
Common near the seashore and on the islands in Salem harbor.

Var. littoralis.
The leaves are narrow.

Var. hastata.
The leaves are halberd-shaped. Both forms are common.

Atriplex arenaria, *Nutt.*
Salisbury, Beverly, Plum Island, Lynn (Tracy). Frequent.

Salicornia herbacea, *L.*
Salt marshes. Common.

Salicornia Virginica, *L.*
Similar places to the last. Common.

Salicornia fruticosa, *L.*, var. **ambigua,** *Gray.*
"Gloucester, Dr. Chas. Pickering, 1824" (Oakes in Hovey's Mag., Vol. XIII); Lynn marsh, 1875 (J. R.).

Suæda maritima, *Dumortier.* Chenopodina of authors. (SEA BLITE.)
Very common on marshes and wet places near the seashore. This plant is so variable that it seems as if more than one species grew here, or at least good varieties might be described. In a letter written July 15, 1875, Dr. Pickering says of this species "Chenopodina Americana not more than a foot high growing in the wet salt marsh and not in the sand, the stem simple or with few branches, root *perennial.*"

Salsola Kali, *L.* (COMMON SALTWORT.)
In sand by the seashore. Common.

AMARANTACEÆ.
(AMARANTH FAMILY.)

Amarantus retroflexus, *L.*
A common weed. (Probably adv. from Trop. Amer.)

Amarantus albus, *L.*
A very common weed. (Probably adv. from Trop. Amer.)
Amarantus caudatus, *L.* (LOVE-LIES-BLEEDING.)
"Amongst rubbish," 1783 (Rev. M. Cutler, Mem. Am. Acad., Vol. I, p. 490). This plant now only seen in old gardens is never noticed as escaped from cultivation.

POLYGONACEÆ.

(BUCKWHEAT FAMILY.)

Polygonum orientale, *L.* (PRINCE'S FEATHER.)
Escaped from cultivation in a few places. Wenham, near the pond; Boxford (Miss Perley), etc. (Adv. from India.)
Polygonum Careyi, *Olney.*
Chebacco woods, in damp places, 1876 (J. R.).
Polygonum Pennsylvanicum, *L.*
Wet, but open places. Not rare.
Polygonum incarnatum, *Ell.*
"In the environs of Salem with P. persicaria" (Dr. Chas. Pickering, Chron. Hist. Pl., p. 904, under P. lapathifolium). Marblehead, 1859 (Rev. J. L. Russell), and damp places elsewhere.
Polygonum Persicaria, *L.* (LADY'S THUMB.)
A weed in waste places. (Nat. from Eu.)
Polygonum Hydropiper, *L.* (WATER-PEPPER.)
Common in wet places as by brooks. "Apparently introduced eastward, but indigenous northward" (Gray's Manual).
Polygonum acre, *H. B. K.* (WATER SMARTWEED.)
Chebacco (J. R.), and a few other places. Rather scarce.
Polygonum hydropiperoides, *Michx.* (MILD WATER-PEPPER.)
In water and wet places. Topsfield (Oakes); Lynn (Tracy); outlet of Chebacco pond; Haverhill, etc.
Polygonum amphibium, *L.,* var. **aquaticum,** *Willd.* (WATER PERSICARIA.)
Floats like a pond weed in the water. Peabody, North Andover, Lynn (Tracy), Andover, Hamilton (G. D. Phippen), etc. Rather frequent.
Var. **terrestre,** *Willd.*
Grows on the shore often near the floating variety.
Polygonum articulatum, *L.* (JOINTWEED.)
Plum Island, Coffin's beach, Gloucester, Ipswich (Oakes), Groveland, Sept. 14, 1859 (Proc. E. I., Vol. II, p. 409).
Var. **multiplex** (Oakes in Hovey's Mag., Vol. VII).
Ipswich, 1825 (Oakes); Turkey hill, Ipswich, 1877 (J. R.). This species resembles a heath and has frequently been collected for Calluna.

Polygonum aviculare, *L.* (KNOTGRASS.)
A very common weed. (Nat. from Eu.)

Polygonum erectum, *L.* (P. aviculare, var. erectum. Gray's Manual.)
Essex County (Rev. J. L. Russell); Ipswich (J. R.). Scarce. Considered an American plant.

Polygonum maritimum, *L.* (COAST KNOTGRASS.)
Beverly, Lynn, Ipswich, etc.

Polygonum tenue, *Michx.*
Andover, Lawrence, Amesbury, Newbury, Lynn (Tracy), Danvers, etc. Frequent.

Polygonum arifolium, *L.* (HALBERD-LEAVED TEAR-THUMB.)
Common in damp places.

Polygonum sagittatum, *L.* (ARROW-LEAVED TEAR-THUMB.)
In similar places. Common.

Polygonum Convolvulus, *L.* (BLACK BINDWEED.)
A climbing weed. (Nat. from Eu.)

Polygonum cilinode, *Michx.*
Danvers, 1840 (memo. Rev. J. L. Russell), Rockport, Andover (Mrs Downs), etc. Not very common.

Polygonum dumetorum, *L.*, **var. scandens.** (CLIMBING FALSE BUCKWHEAT.)
Beverly, West Gloucester, Lynn (Tracy), Manchester (G. D. Phippen), etc. Frequent.

Fagopyrum esculentum, *Mœnch.* (BUCKWHEAT.)
Frequent in cultivation, occasionally escaping, Lynn (Tracy), Danvers (J. H. Sears). (Adv. from Eu.)

Rumex orbiculatus, *Gray.* (GREAT WATER-DOCK.)
Essex County (Oakes), West Newbury, Swampscott. Scarce.

Rumex salicifolius, *Weinmann.* (WHITE DOCK.)
Nahant (Tracy), Gloucester, Newburyport, etc., near the seashore. "First sent by Dr. Nichols from Danvers" (Bigelow's Fl. Bost., 2d ed., 1824).

Rumex crispus, *L.* (CURLED DOCK.)
A roadside weed. (Nat. from Eu.)

Rumex obtusifolius, *L.* (BITTER DOCK.)
A common weed. (Nat. from Eu.)

Rumex Acetosella, *L.* (FIELD OR SHEEP SORREL.)
Another ubiquitous weed. Noticed near Salem (1628) by Higginson and by some authors considered an American plant. Mr. Phippen mentions that strong plants in the garden bear tubers on the roots. (Nat. from Eu.)

LAURACEÆ.

(LAUREL FAMILY.)

Sassafras officinale, *Nees.* (SASSAFRAS.)
Frequent in most towns. Old and large trees are becoming rare.

Lindera Benzoin, *Meisner.* (SPICE-BUSH; BENJAMIN-BUSH.)
Moist woods. Frequent.

THYMELEACEÆ.

(MEZEREUM FAMILY.)

Daphne Mezereum, *L.*
Frequently cultivated, and found escaped at Ipswich, in 1877, by Dr. Chas. Palmer; and, at Salem, 1879 (Harry H. Patch).

SANTALACEÆ.

(SANDALWOOD FAMILY.)

Comandra umbellata, *Nutt.* (COMANDRA.)
Frequent in rather open places.

CERATOPHYLLACEÆ.

(HORNWORT FAMILY.)

Ceratophyllum demersum, *L.*
Ponds and streams. Frequent. The only specimens found in fruit have been collected in Pleasant pond, Wenham.

CALLITRICHACEÆ.

(WATER-STARWORTS.)

Callitriche verna, *L.*
Ponds and streams. Very common. Found fruiting while only half an inch high when growing in mud, and also in deep water where it frequently grows more than a foot long. C. heterophylla, *Pursh,* and C. autumnalis, *L.*, have been reported as growing in the county, the former in Lynn by Tracy, the latter in Groveland by Russell, but neither are in the county collection of the P. A. S. That there should be confusion regarding these plants is not at all surprising, as the terrestrial and aquatic forms are so unlike; and it may be that there are two species to be found here.

EUPHORBIACEÆ.

(SPURGE FAMILY.)

Euphorbia polygonifolia, *L.*
Beaches, in sand, from Nahant to Salisbury, and Plum Island.

Euphorbia maculata, *L.*
Dry places, etc. A common weed.

Euphorbia hypericifolia, *L.*
"Kenoza," Haverhill, 1875 (J. R.); and one or two other localities.

Euphorbia marginata, *Pursh.*
A native of Kansas and Nebraska. Frequently cultivated, escaping from gardens occasionally.

Euphorbia Esula, *L.*
Very abundant near Newburyport on the line of the Eastern R. R. (Nat. from Eu.) This is probably Oakes's locality. The E. Esula in Tracy's Lynn Flora is probably the next species which is the common one in the southern portion of the county.

Euphorbia Cyparissias, *L.*
Very common in the older places where it has escaped from gardens. (Nat. from Eu.)

Acalypha Virginica, *L.* (THREE-SEEDED MERCURY.)
"Yards and rubbish heaps in Lynn (Tracy); Chebacco, Boxford, etc. Frequently varies greatly in size, small forms being only a few inches high, while others are quite fifteen.

URTICACEÆ.

(NETTLE FAMILY.)

Ulmus fulva, *Michx.* (SLIPPERY ELM.)
Boxford (J. H. Sears), Georgetown (Mrs. Horner), Wenham. Apparently indigenous, although scarce.

Ulmus Americana, *L.* (AMERICAN OR WHITE ELM.)
Common as a shade tree and very freely reproduced by seeds.

Celtis occidentalis, *L.* (NETTLE-TREE; HACKBERRY.)
The only specimens of this species, now known to the writer, are at the summit of one of the dunes on Plum Island, and between Salem and Lynn in the Great Pastures. There are specimens from the county in the P. A. S. herbarium from Ipswich collected by the late Mr. Oakes, and Mr. S. P. Fowler remembers a tree in Topsfield, which has since been cut down. There are also four trees of this species on the farm of Mr. Towns, at Boxford, which were raised from seeds obtained at Lowell, about 1840.

Morus rubra, *L.*
"In Essex County, Mass. (Oakes)" (De Candolle's Prodromus XVII, p. 246). The specimens in the collection of the P. A. S. from Oakes are marked "New England," probably from western Massachusetts or Vermont.

M. alba and M. nigra.
Oriental species, having long been in cultivation, are to be found near old estates and in gardens. They are rarely seen, however, thoroughly escaped.

Urtica dioica, *L.* (COMMON NETTLE.)
Fence rows, etc. Very common. (Nat. from Eu.)

Urtica gracilis, *Ait.*
Somewhat resembling the last, and found frequently in similar situations. A native plant.

Urtica urens, *L.* (SMALL NETTLE.)
"Occasional in Lynn" (Tracy), Rockport (Pool); West Gloucester, Hamilton, etc. Not very common. (Nat. from Eu.)

Laportea Canadensis, *Gaudichaud.* (WOOD-NETTLE.)
This appeared in Salem in the garden of Rev. E. C. Bolles and that of the writer within a few years and continues. They undoubtedly came from seeds in the earth, about plants brought from the woods, but from what locality it is not known.

Pilea pumila, *Gray.* (RICHWEED; CLEARWEED.)
Common in damp shady places.

Bœhmeria cylindrica, *Willd.* (FALSE NETTLE.)
Beaver pond region, Beverly; Essex, Georgetown, Lynn (Tracy), Amesbury, Haverhill, etc. Occasional in wet places.

Cannabis sativa, *L.* (HEMP.)
"Occasional in Lynn" (Tracy); Rockport (Tracy, 1863, Proc. E. I. Vol. III, p. 276); Beverly, 1875; Salem (G. D. Phippen); Newburyport (J. R.). Not seemingly permanent in any one place, although frequently met with in the county. (Adv. from Eu.)

Humulus Lupulus, *L.* (HOP.)
"Indigenous northward and westward" (Gray's Manual). Much cultivated and frequently found escaped.

PLATANACEÆ.

(PLANE-TREE FAMILY.)

Platanus occidentalis, *L.* (BUTTONWOOD.)
Frequent, both in cultivation and also in remote places. "Ob-

served by Josselyn, voyage, 1670, "a stately tree growing here and there in the valleys." "A buttonwood tree, which measured nine yards in girth," is mentioned by Paul Dudley, writing from New England in 1726 (Phil. Trans., XXXIII, 129, etc.). Not clearly indigenous along the Atlantic in New England" (Dr. Chas. Pickering, Chron. Hist. Pl. p. 961). Probably not a native of Essex county, but introduced from the westward, as suggested by Dr. Pickering.

JUGLANDACEÆ.

(WALNUT FAMILY.)

Juglans cinerea, *L.* (BUTTERNUT.)
Common along walls but not so in forests. Tracy considers it an introduced tree in Lynn.

Juglans nigra, *L.* (BLACK WALNUT.)
Occasionally cultivated from the western states. We have received specimens of the nuts of this tree from Mr. J. C. Peabody of Newburyport, gathered in a place near that city where he thinks the trees are indigenous, but this seems under careful consideration to be hardly probable.

Juglans regia, *L.* (EUROPEAN WALNUT.)
Cultivated in Salem, to a small extent, where the fruit perfects well. The J. regia, noticed in Saugus (Proc. E. I., Vol. II, p. 389), proves to be J. nigra.

Carya alba, *Nutt.* (SHAG-BARK HICKORY.)
Common in most towns.

Carya tomentosa, *Nutt.* (MOCKER-NUT; WHITE-HEART HICKORY.)
Danvers (J. H. Sears). Scarce. This tree is more abundant to the south of Boston than in Essex county.

Carya porcina, *Nutt.* (PIG-NUT HICKORY.)
Common in most parts of the county.

Carya amara, *Nutt.* (BITTER-NUT OR SWAMP HICKORY.)
Quite common in most towns. Particularly fine trees are to be found in Boxford with leaves much more delicate in cutting than the usual form, giving the tree a different appearance.

CUPULIFERÆ.

(OAK FAMILY.)

Quercus alba, *L.* (WHITE OAK.)
Common throughout the county; occasionally fifteen to twenty feet in circumference.

Quercus Robur, *L.* (ENGLISH OAK.)
Cultivated often, but a tree of slow growth. A few small trees are growing on the roadside of Salem turnpike having escaped from the Fay estate where there are fine specimens of this species.
Var. **pedunculata** seems to be the most common form here.

Quercus bicolor, *Willd.* (SWAMP WHITE OAK.)
Common, mostly in wet places or low grounds. Very variable both as to the whiteness on the under side of the leaves, their cutting, and the mossy fringe of the cup.

Quercus Prinus, *L.* (CHESTNUT OAK.)
The region of Topsfield, Boxford, Georgetown, Middleton, and North Andover, seems to be the only locality from which this species is collected. One tree noticed in Boxford differs much from the ordinary form and may prove to be another variety. Some forms quite closely resemble the larger specimens of the next species.

Quercus prinoides, *Willd.* (CHINQUAPIN-OAK.)
"South Danvers (Peabody) Poor Farm, June 5, 1857" (S. P. Fowler, Proc. E. I., Vol. II, p. 204); Georgetown (Mrs. Horner); Boxford and Topsfield, frequent. Absent in the southern and eastern portions of the county.

Quercus ilicifolia, *Wang.* (BEAR OR BLACK SCRUB-OAK.)
Common in poor soil; most abundant in Peabody, Lynnfield and Topsfield. In some towns absent. There is a variety with more depressed acorns having rougher cups and seeming different from the common form.

Quercus coccinea, *Wang.* (SCARLET OAK.)
Boxford, Andover, Middleton, Topsfield, Danvers. The fine and deep cutting of the leaves gives the trees an exceedingly graceful appearance. This species is quite distinct from the next and does not seem to be present in many towns.

Quercus tinctoria, *Bart.* (BLACK OAK; YELLOW-BARKED OAK.)
Abundant. Easily distinguished from the preceding by the general coarser appearance and the yellow inner bark.

Quercus rubra, *L.* (RED OAK.)
Common in all parts of the county.

Castanea vulgaris, *Lam.,* var. **Americana,** *A. DC.* (CHESTNUT.)
Castanea vesca, *L.* (Gray's Manual.)
Lynnfield, Danvers, Lynn (Tracy), towns on the Merrimac, etc. Not very common.

Fagus ferruginea, *Ait.* (AMERICAN BEECH.)
"Rare in Lynn" (Tracy), Chebacco abundant, Merrimac valley, etc.

Corylus Americana, *Walt.* (COMMON HAZEL-NUT.)
Along walls and roadsides. Common.
Corylus rostrata, *Ait.* (BEAKED HAZEL-NUT.)
Similar places, but not so common.
Ostrya Virginica, *Willd.* (HOP-HORNBEAM.)
Frequent in many towns.
Carpinus Caroliniana, *Walt.*, Carpinus Americana, *Michx.* (Gray's Manual.) (HORNBEAM; BLUE BEECH.)
Rather less common than the last.

MYRICACEÆ.

(SWEET-GALE FAMILY.)

Myrica Gale, *L.* (SWEET GALE.)
Swamps and borders of ponds. Common.
Myrica cerifera, *L.* (BAYBERRY.)
Pastures, etc. Common.

Comptonia asplenifolia, *Ait.* (SWEET-FERN.)
Pastures and hills. Common.

BETULACEÆ.

(BIRCH FAMILY.)

Betula lenta, *L.* (SWEET OR BLACK BIRCH.)
Frequent.
Betula lutea, *Michx.*, f. (YELLOW OR GRAY BIRCH.)
Quite common.
Betula alba, var. **populifolia,** *Spach.* (WHITE BIRCH.)
Very common on hillsides, and by the roads.
Betula papyracea, *Ait.* (PAPER OR CANOE BIRCH.)
Occasional. Salem Great Pastures. Wenham pond, north shore. Danvers, and more frequently in the northwestern portion of the county.
Betula nigra, *L.* (RED OR RIVER BIRCH.)
North Andover and Lawrence, near the Merrimac river.
Betula pumila, *L.* (LOW BIRCH.)
"Shore of North Andover pond" and "on a railroad track in Groveland" (S. P. Fowler, Proc. E. I., Vol. II, pp. 402 and 409), Rockport (Pool, in "Pigeon Cove and vicinity"). Specimens brought to the P. A. S., in 1879, from the shore of Pentucket pond,

Georgetown, as this species, are but small specimens of B. nigra. Not represented in the county herbarium. Very doubtful as a native of this region.

Alnus incana, *Willd.* (SPECKLED OR HOARY ALDER.)
Frequent in wet places.

Alnus serrulata, *Ait.* (COMMON ALDER.)
By streams and moist roadsides. Abundant.

SALICACEÆ.

(WILLOW FAMILY.)

Salix candida, *Willd.* (HOARY WILLOW.)
A meadow in Boxford, the first locality reported in this region, plants of both sexes being quite numerous. First noticed Aug. 1875 (J. R.).

Salix tristis, *Ait.* (DWARF GRAY WILLOW.)
Lynnfield, Danvers (J. H. Sears), etc. In places occupied by the Bear Oak. Not very common.

Var. **microphylla.** (Gray's Manual.)
Essex County (Oakes).

Salix humilis, *Marshall.* (PRAIRIE WILLOW.)
Dry places. Rather common. Many of the willows have cones, persistent after the leaves fall, upon the ends of the branches. They are caused by the sting of an insect; and, if cut into when green, the larva of the fly will be found within.

Salix discolor, *Muhl.* (PUSSY WILLOW).
Common in damp places.

Salix sericea, *Marshall.* (SILKY WILLOW.)
Along streams. Not very common.

Salix petiolaris, *Smith.*
Middleton, Andover, Methuen, Chebacco, etc. Frequent.

Salix purpurea, *L.*
Swampscott, Salem, Newburyport. Introduced for basket work. (Adv. from Eu.)

Salix viminalis, *L.* (BASKET OSIER.)
Danvers, 1853 (Dr. Osgood's list). Reported in Salem and one or two other localities, but it can hardly be called a thoroughly naturalized plant. Bigelow (Fl. Bost., 2d. ed., 1824) speaks of this species as growing "In swamps at Danvers and elsewhere." As he does not mention Salix sericea, it may be this species that is referred to, as it is frequent in swamps in the vicinity of Danvers, the leaves somewhat resembling those of S. viminalis. (Adv. from Eu.)

Salix cordata, *Muhl.*
Wet places. Frequent. Quite variable.
Salix livida, *Wahl,* var. **occidentalis.** (Gray's Manual.)
Quite common in rather dry places. A very curious abnormal form of this species was collected in Beverly, August, 1879, by Mr. Sears. The plant had for some reason put forth a new growth having catkins at the end of very short leaf-bearing branches. The bracts of the catkins were large and green, and the yellow stamens very distinct.
Salix lucida, *Muhl.* (SHINING WILLOW.)
By streams and ponds. Frequent.
Salix nigra, *Marsh,* var. **falcata.** (Gray's Manual.)
Topsfield, Danvers, Chebacco, Andover, Essex, etc. Frequent. The largest of our native willows.
Salix fragilis, *L.*
" On the turnpike road to Salem " (Emerson, Trees and Shrubs of Mass.), Ipswich (Oakes). (Adv. from Eu.)
Salix alba, *L.*
The common large willow in Essex county. There are several varieties of this species. (Adv. from Eu.)
Var. **vitellina** (Gray's Manual).
The most beautiful, although not so common as the ordinary form.
Salix Babylonica, *Tourn.* (WEEPING WILLOW.)
Common only under cultivation.
Salix myrtilloides, *L.*
Topsfield (Oakes), Essex, Boxford, Danvers (J. H. Sears). In cold bogs. Rather scarce.
Populus tremuloides, *Michx.* (AMERICAN ASPEN.)
Various parts of the county. Common.
Populus grandidentata, *Michx.*
Similar places to the last. Common.
Populus balsamifera, *L.,* var. **candicans.** (BALM OF GILEAD.)
A native of northern New England. Much planted and spreading by root suckers.
Populus dilatata, *Ait.* (LOMBARDY POPLAR.)
Once a common shade or ornamental tree, but fast disappearing. (Adv. from Eu.)
Populus alba, *L.*
Common in cultivation where it spreads much. (Adv. from Eu.)

GYMNOSPERMS.

CONIFERÆ.

(PINE FAMILY.)

Pinus rigida, *Miller.* (PITCH PINE.)
A common tree.

Pinus resinosa, *Ait.* (RED, OR NORWAY PINE.)
A grove in Boxford; a few trees in Georgetown (Mrs. Horner); West Newbury (Wm. Merrill); one tree in Andover (Prof. Goldsmith); two trees in Peabody (Mr. Brown). Our rarest conifer.

Pinus Strobus, *L.* (WHITE PINE.)
The forest tree of Essex county at the present time.

Pinus Larico, *L.*, var. **Austriaca,** *Engl.* (AUSTRIAN PINE), and its cogener and frequent companion,

Pinus sylvestris, *L.* (SCOTCH PINE), have long been introduced from Europe, and are much planted as ornamental trees. The latter has spread by seeds to a small extent in Danvers.

Picea nigra, *Link.*, Abies nigra, *Poir.* (Gray's Manual). (BLACK SPRUCE.)
Frequent in the Essex, Chebacco, and Middleton woods, and Pine Swamp, at Ipswich; also in a few other places. Rare in most towns. The Picea alba (White Spruce) and Abies balsamea (Balsam Fir) are only found in cultivation.

Tsuga Canadensis, *Cariere*, Abies Canadensis, *Michx.* (Gray's Manual). (HEMLOCK SPRUCE.)
Frequent in almost every town.

Larix Americana, *Michx.* (LARCH; TAMARACK.)
Wenham, Boxford, Ipswich, Lynn (Tracy). Becoming scarce. The European larch flourishes better in cultivation (as does the Norway spruce) than the corresponding American species.

Chamæcyparis sphæroidea, *Spach*, Cupressus thyoides, *L.* (Gray's Manual.) (WHITE CEDAR.)
Essex county has been considered to be about the northern limit of this species in this region, but the writer feels quite sure that it probably extends to Portsmouth, N. H., as a swamp near Greenland, near Portsmouth, on the line of the E. R. R., apparently contains

an abundance of this species of tree. The so-called White Cedar of Maine and New Hampshire is the Thuja occidentalis (Arbor vitæ), which is here only known in cultivation. Taxodium distichum (Bald Cypress) is cultivated in a few instances, proving hardy in Salem where there are a few large trees.

Juniperus communis, *L.* (JUNIPER.)
Sterile hillsides. Common. This is very frequently called Savin, which name belongs to the next species.

Juniperus Virginiana, *L.* (RED CEDAR.)
Common on dry hills.

Taxus baccata, *L.*, var. **Canadensis,** *Gray.* (AMERICAN YEW; GROUND HEMLOCK.)
Rockport, Danvers, Essex, Beverly, etc. Rather scarce. The yew of Europe is a tree rarely cultivated here; the only specimen now called to mind being in Salem, in the garden of Mr. Chas. M. Richardson.

ENDOGENS.

ARACEÆ.

(Arum Family.)

Arisæma triphyllum, *Torrey.* (Indian Turnip; Jack-in-the-pulpit.)
Damp shady places. Common. In cultivation, it will often grow over three feet high.

Peltandra Virginica, *Raf.* (Arrow Album.)
Bogs, etc. Frequent.

Calla palustris, *L.* (Wild Calla.)
In very wet muddy places. Rather scarce.

Symplocarpus fœtidus, *Salisb.* (Skunk Cabbage.)
Common in wet places.

Acorus Calamus, *L.* (Sweet Flag.)
Meadows and brooks. Frequent.

LEMNACEÆ.

(Duckweed Family.)

Lemna polyrrhiza, *L.*
Stagnant water. Frequent.

Lemna minor, *L.*
Vicinity of Danvers (Dr. Osgood's list), West Gloucester, etc. Probably not rare.

TYPHACEÆ.

(Cat-tail Family.)

Typha latifolia, *L.* (Common Cat-tail.)
Brooks, etc. Common.

Typha angustifolia, *L.*
Rockport(?), "Lynn marshes, two fertile spikes" (memo. Dr. Chas. Pickering). Specimens in the hands of a small boy were obtained by Miss Brooks, 1880, probably collected in the vicinity of Salem.

Sparganium eurycarpum, *Engelm.* (BUR-REED.)
"1823 * * my first botanical discovery * * * observed in the extensive marsh between Ipswich river and Wenham swamp; communicated to Nuttall, but it did not attract his attention " (Dr. Chas. Pickering, Chron. Hist. Pl., p. 1063). Common in slow waters.

Sparganium simplex, *Huds.*
"Outlet of Pleasant pond, Wenham, about 1860" (Dr. Chas. Pickering).

Var. **Nuttallii,** *Engelm.*
Foster's pond, North Beverly.

Var. **angustifolium,** *Engelm.*
Merrimac river, at Lawrence; Crane's pond, West Newbury. Rare.

NAIADACEÆ.

NOTE.—This order contains many species which are very difficult to determine, and which require close observation of living specimens in order that they may be correctly placed. Mr. Charles E. Faxon, of Boston, who has carefully studied these plants from specimens collected in the stations mentioned below, nearly all of which he has himself visited, kindly revised the original list, adding many notes of value.

(POND WEED FAMILY.)

Naias flexilis, *Rostk.*
Common in brooks and ponds.

Naias Indica, var. **gracillima,** *Braun.*
In a springy bog at Sluice pond, Lynn, 1880 (Edwin Faxon). This species is occasionally met with in similar places, but no other authentic specimens from Essex county have been found.

Zannichellia palustris, *L.* (HORNED PONDWEED.)
In the brook or "upper branch" of the Mill pond above the Atlantic Car Works, Salem, in brackish water.

Zostera marina, *L.* (EEL-GRASS.)
Everywhere in salt water, below low tide mark, along the shore.

Ruppia maritima, *L.*
In shallow, salt or brackish water, inlets, etc. Frequent.

Potamogeton natans, *L.*
Ponds and rivers. Common.

Var. **prolixus,** *Koch.*
In flowing water, outlet of Chebacco pond, Ipswich river, etc. Frequent.

Potamogeton Oakesianus, *Robbins.*
Foster's pond, Beverly, 1873 (J. R.); Sluice pond, Lynn (Edwin Faxon).
Potamogeton Claytonii, *Tuckerman.*
Ponds and brooks. Common.
Potamogeton Spirillus, *Tuckerman.*
Old mill brook, Boxford; Merrimac river at Lawrence, Georgetown. There is a variety, with no floating leaves, in Chebacco pond.
Potamogeton hybridus, *Michx.*
Common in stagnant water.
Potamogeton lonchites, *Tuckerman.*
Ipswich river. Merrimac river at Lawrence.
Potamogeton pulcher, *Tuckerman.*
Chebacco pond; Ipswich river; Beaver pond, Beverly; Crane pond, West Newbury.
Potamogeton amplifolius, *Tuckerman.*
Wenham pond, Ipswich river, ponds in Andover and Haverhill; fine fruit Aug. 19 (1879), Crane pond, West Newbury, river Parker, etc.
Potamogeton gramineus, *L.*
Wenham, and Pleasant ponds, Wenham. Quite abundant and variable.
Var. **heterophyllus,** *Fries.*
Georgetown (J. H. Sears), Sluice pond (C. E. Faxon).
Potamogeton lucens, *L.*
"Breed's pond, Lynn" (Tracy). Recent search has failed to verify this determination; and specimens, which at first were supposed to be this species, proved to be imperfectly developed forms of P. amplifolius.
Var. **minor,** *Nolte.*
Pleasant pond, Wenham (J. R., etc.). Scarce. This will probably prove a good species, and as it is the same as the European P. Zizii, *M. & K.*, it will be given under that name.
Potamogeton prælongus, *Wulfen.*
Wenham and Pleasant ponds, Wenham.
Potamogeton perfoliatus, *L.*
Ponds and streams. Frequent.
Potamogeton zosteræfolius, *Schumacher,* Potamogeton compressus, *L.* (Gray's Manual).
River Parker, West Newbury (J. R.); Lynnfield (Edwin Faxon). Rare.
Potamogeton obtusifolius, *Mertens & Koch.*
Pleasant pond, Wenham, Aug., 1875 (J. R.), 1880 (C. E. Faxon).
Potamogeton pauciflorus, *Pursh.*
Wenham pond; Flax pond, Lynn (Tracy), Boxford.

Potamogeton pusillus, *L.*
Marblehead (Rev. J. L. Russell), Danversport, in brackish water; outlets of Wenham and Chebacco ponds.

Var. tenuissimus, *Mertens & Koch.*
Wenham pond.

Potamogeton gemmiparus, *Robbins.* Potamogeton pusillus, var. (?) gemmiparus. (Gray's Manual).
Wenham pond.

Potamogeton Robbinsii, *Oakes.*
"Wenham pond, 1829" (Oakes in Hovey's Mag., Vol. VII), the original discovery; Crane pond, West Newbury; Sluice pond, and abundant in Flax pond (C. E. Faxon). Single specimens have been noticed at Pleasant pond, Wenham, and Ipswich river. It varies to small forms.

Potamogeton crispus, *L.*
Spy pond, Arlington, 1880 (C. E. Faxon). Probably introduced.

ALISMACEÆ.

(WATER-PLANTAIN FAMILY.)

Triglochin maritimum, *L.*
Salt marshes. Common.

Scheuchzeria palustris, *L.*
"Wenham swamp, 1824" (memo. Dr. Chas. Pickering).

Alisma Plantago, *L.*, var. **Americanum.**
Wet places along roadsides and railroad beds. Common.

Echinodorus parvulus, *Engelm.*
Near Cambridge (Prof. James); Winter pond, Winchester (E. H. Hitchings). This little plant ought to be found within the limits of the county as it occurs so near us.

Sagittaria variabilis, *Engelm.*
A plant of many forms.

Var. obtusa.
Georgetown (Mrs. Horner); Essex, etc. Quite common.

Var. latifolia.
Brooks and meadows. Frequent.

Var. hastata.
Ditches, etc. This is the most common form.

Var. diversifolia.
Pleasant pond, Wenham.

Var. angustifolia.
Pleasant pond, Wenham.
Var. gracilis.
Georgetown (Mrs. Horner).
Sagittaria heterophylla, *Pursh.*
Shore of the Merrimac, Lawrence, two miles above the dam, north side, Aug. 5, 1879 (J. R., etc.).
Sagittaria graminea, *Michx.*
Borders of Wenham pond (J. R.).
Forma fluitans, *Engel.*
With leaves two feet long collected in the Ipswich river by Mr. Sears. Similar specimens were found near Boston by Mr. E. H. Hitchings.

HYDROCHARIDACEÆ.

(FROG'S-BIT FAMILY.)

Anacharis Canadensis, *Planchon.* (WATER-WEED.)
Very common in some places, and undoubtedly to be found in the county, but no specimens are in the P. A. S. collection.
Vallisneria spiralis, *L.* (VALLISNERIA.)
Frequent in ponds. A most interesting plant to study, the mode of fertilization being very curious.

ORCHIDACEÆ.

(ORCHIS FAMILY.)

Orchis spectabilis, *L.* (SHOWY ORCHIS.)
Not known within the exact geographical limits of the county, but growing quite near the line toward Chelsea, yet Dr. Bigelow who botanized in that region does not mention it.
Habenaria tridentata, *Hook.*
Essex, Beverly, Lynnfield (A. P. Chute), Salem Great Pastures (Tracy), Georgetown, Amesbury, North Andover. Scarce.
Habenaria virescens, *Spreng.*
Along brooks and in meadows, Georgetown (Mrs. Horner), etc. Not very common. It is quite probable that the species referred to as O. dilatata, Proc. E. I. Vol. II, p. 240, is either this or H. lacera.
Habenaria Hookeri, *Torrey.*
Found at Georgetown, June, 1874, by Mrs. Horner. It was exhibited among other native plants in Boston, at the Mass. Hort. Soc. rooms, at that date.

Habenaria orbiculata, *Torrey.*
"In Danvers, Dr. Nichols" (Bigelow's Fl. Bost. 2d. ed., 1824), "In the deep woods at Gloucester" (Mrs. Bray and Mrs. Grover) where the only specimen found was a very fine one.

Habenaria ciliaris, *R. Br.* (YELLOW-FRINGED ORCHIS.)
"Lynnfield, 1856" (Proc. E. I., Vol. II, p. 47). This proves to be a mistake, another plant being intended, as probably is also the "Orchis ciliaris" of Dr. Osgood's list.

Habenaria blephariglottis, *Hook.* (WHITE-FRINGED ORCHIS.)
"Great Swamp, Amesbury" (J. G. Whittier), Lynnfield (Rev. A. P. Chute), Rockport (Pool), Essex County, 1824 (Oakes and Pickering), Magnolia swamp, Gloucester (John C. Phillips). Scarce.

Habenaria lacera, *R. Br.*
Frequent in meadows.

Habenaria psycodes, *Gray.* (PURPLE ORCHIS.)
Near brooks and in meadows. Seldom abundant in any locality yet found in many places.

Habenaria fimbriata, *R. Br.* (GREAT PURPLE ORCHIS.)
Ipswich (Oakes); Andover (Mrs. Downs); Danvers (J. H. Sears); Georgetown (Mrs. Horner). Not so frequent as the last.

Goodyera repens, *R. Br.*
Pine woods. Not very common.

Goodyera pubescens, *R. Br.*
Similar situations to the last, and more frequent, flowering later.

Spiranthes cernua, *Richard.* (LADIES' TRESSES.)
Common in low grounds in September.

Spiranthes latifolia, *Torr.*
Lynnfield, June, 17, 1879 (Herbert A. Young). Rare.

Spiranthes gracilis, *Bigelow.*
Common in rather dry pastures.

Listera cordata, *R. Brown.*
Magnolia, May, 1880 (Chas. J. Sprague). Rare.

Arethusa bulbosa, *L.* (ARETHUSA.)
Frequent in meadows and bogs. Usually rose-purple but sometimes pure white. One specimen from Hamilton (Caleb Cooke), 1879, had a perfect flower completely inverted. Monstrosities are also found having two flowers on one stalk.

Pogonia ophioglossoides, *Nutt.* (POGONIA.)
Frequent in meadows and shores of ponds. Rose color to white. In cool swamps the flowers are found as late as August.

Pogonia verticillata, *Nutt.*
North Andover (Prof. Goldsmith), Haverhill (E. H. Hitchings). More frequent south of Boston.

Calopogon pulchellus, *R. Br.* (CALOPOGON.)
Meadows. Not rare.

Microstylis ophioglossoides, *Nutt.* (ADDER'S-MOUTH.)
"Manchester, Oakes and Osgood" (memo. Rev J. L. Russell), Essex County (Dr. Chas. Pickering), Wilmington (E. H. Hitchings).

Liparis Lœselii, *Richard.* (TWAY BLADE.)
Beverly, 1872 (J. R.), Andover (Rev. H. P. Nichols), "very rare in Lynn, Dr. Holder" (Tracy), Topsfield, Lynnfield (Rev. A. P. Chute). Rare.

Corallorhiza innata, *R. Brown.*
Manchester (Oakes), Magnolia Swamp (J. H. Sears and J. R.), Chebacco (Miss. L. H. Upton). Rare.

Corallorhiza odontorhiza, *Nutt.*
"Rare. Rather plentiful on the east side of Edwards' Swamp" (Tracy). Not represented in the P. A. S. County herbarium.

Corallorhiza multiflora, *Nutt.*
Chebacco, Rockport (Pool), Georgetown (Mrs. Horner), "Rare. Blood Swamp hills" (Tracy). More frequent than the others, yet scarce.

Aplectrum hyemale, *Nutt.* (ADAM AND EVE; PUTTY-ROOT.)
Reported by Mr. G. D. Phippen as collected at a field meeting of the Essex Institute, held at West Lynn, Aug. 26, 1857 (Proc. E. I. Vol. II, p. 225); also stated by Mrs. Horner to have been found at Georgetown some years since. Not represented in the P. A. S. county herbarium.

Cypripedium parviflorum, *Salisb.* (SMALLER YELLOW LADY'S SLIPPER.)
Danvers (J. H. Sears). Scarce.

Cypripedium pubescens, *Willd.* (LARGER YELLOW LADY'S SLIPPER.)
Reported from Wenham and Haverhill in private plant-lists; but as no specimens have been preserved, it is quite probable that those found should be referred to the last species.

Cypripedium spectabile, *Swartz.*
Found in 1877, by J. H. Sears, in an unfrequented part of the great Wenham swamp. Several other plants, not common in the county, and often associated with this species in places where it is more abundant, are found with it here. It was also found at Andover (Jackson T. Dawson). This fact seems to prove its claim to be an Essex County plant.

Cypripedium acaule, *Ait.* (COMMON LADY'S SLIPPER.)
> Common in rather dry woods, chiefly under conifers.

Var. alba.
> Leaves lighter green, flowers large, pure white. There are light colored flowers on the ordinary plants, but the flowers of the variety are whiter still. Georgetown (Mrs. Horner), Beverly (G. D. Phippen), N. Reading. Scarce.

AMARYLLIDACEÆ.

(AMARYLLIS FAMILY.)

Hypoxys erecta, *L.* (STAR-GRASS.)
> Damp places, among other plants. Common.

IRIDACEÆ.

(IRIS FAMILY.)

Iris versicolor, *L.* (BLUE FLAG.)
> Common in meadows and by brooks.

Iris Virginica, *L.* (SLENDER BLUE FLAG.)
> "Stony brook, Johnson's field, Lynn" (Tracy), Andover abundant, Ipswich (Oakes), etc. Not so common as the last except in the vicinity of Andover where it is the prevailing species.

Sisyrinchium Bermudiana, *L.* (BLUE-EYED GRASS.)
> Fields, etc. Common.

SMILACEÆ.

(SMILAX FAMILY.)

Smilax rotundifolia, *L.* (GREENBRIER.)
> Thickets. Common.

Smilax herbacea, *L.* (CARRION FLOWER.)
> Roadsides and moist fields. Frequent.

LILIACEÆ.

(LILY FAMILY.)

Trillium erectum, *L.* (PURPLE TRILLIUM.)
> Blind Hole Swamp, Danvers (J. H. Sears). Along the Ipswich river, Swampscott, 1879 (W. P. Andrews). "Saw two plants in Swampscott May 12, 1868" (memo. Rev. J. L. Russell). Haverhill (Mrs. Downs). Occasional.

Trillium cernuum, *L.* (NODDING TRILLIUM.)
Frequent in damp shady places.
Trillium erythrocarpum, *Michx.* (PAINTED TRILLIUM.)
Manchester (Mrs. Babson); "Amesbury" (Mrs. Downs). Rare.
Medeola Virginica, *L.* INDIAN CUCUMBER-ROOT.)
Common in damp shady places.
Veratrum viride, *Ait.* (AMERICAN WHITE HELLEBORE; INDIAN POKE).
Frequent in swamps and along brooks.
Uvularia perfoliata, *L.*
Frequent under hard-wood trees.
Oakesia sessilifolia, *Watson,* Uvularia sessilifolia, *L.* (Gray's Manual.) (COMMON BELLWORT.)
In open woods. Common. It is a great pleasure to be able in this, the first general enumeration of the plants of Essex county, to introduce the new name given to one of our most abundant and prettiest spring flowers, the Bellwort. Prof. Sereno Watson in "Contributions to American Botany, IX, Revision of the North American Liliaceæ" (Proceedings of the American Academy of Arts and Sciences, Vol. XIV), says:—"The division of Uvularia itself, which seems to be required, affords an opportunity to honor the memory of the lamented botanist, Mr. William Oakes, whose persistent zeal in investigating the flora of the fields and mountains of his native New England, makes appropriate the union of his name with one of the plants, which he himself knew so well." The writer is also indebted to Professor Watson for much valuable assistance in determining many plants.
Streptopus roseus, *Michx.* (TWISTED STALK.)
Rockport (Pool); Gloucester (Mrs. Babson). Rare.
Clintonia borealis, *L.* (CLINTONIA.)
In old deep woods. Not very common.
Convallaria majalis, *Raf.* (LILY OF THE VALLEY.)
This plant is found in Virginia in a wild state, although the species as cultivated here was undoubtedly introduced from Europe. The two plants are considered by Gray as identical. Old gardens, escaping into walks and roadsides. Cultivated everywhere.
Smilacina racemosa, *Desf.* (FALSE SPIKENARD.)
Along walls in shady places. Common.
Smilacina stellata, *Desf.*
Moist places. Most common near the coast.
Smilacina trifolia, *Desf.*
Gloucester and Rockport. Scarce.

Maianthemum Canadense, *Desf.* Smilacina bifolia, var. Canadensis, *Ker.* (Gray's Manual). (WILD LILY OF THE VALLEY.)
Common in all the pine woods. This species has been restored by Prof. Watson in his "Revision of the Liliaceæ" to Maianthemum.

Polygonatum biflorum, *Ell.* (SMALLER SOLOMON'S SEAL.)
Shady places near walls.

Polygonatum giganteum, *Dietrich.* (GREAT SOLOMON'S SEAL.)
"Ship rock," G. D. Phippen (Proc. E. I., Vol. II, p. 204); "Haverhill," Mrs. S. M. Downs. Not represented in the collection of the P. A. S., and possibly only large forms of the last.

Asparagus officinalis, *L.* (GARDEN ASPARAGUS.)
Often found in fields and roadsides where seedlings have obtained a foothold. (Adv. from Eu.)

Lilium Philadelphicum, *L.* (WILD ORANGE-RED LILY.)
Frequent in open, rather dry, places.

Lilium Canadense, *L.* (WILD YELLOW LILY.)
Meadows; not so abundant as the last. This species is said to succeed well under cultivation.

Lilium superbum, *L.* (TURK'S-CAP LILY.)
"Danvers" (J. H. Sears). The only locality yet reported for the county.

Lilium tigrinum, *L.* (TIGER LILY.)
Escaped in a few places in Wenham, Byfield, Danvers, etc. (Nat. from Eu.)

Erythronium Americanum, *Smith.*
In moist ground under hard wood trees. Very abundant. The common name of this plant, *Dog's-tooth Violet,* is absurd. *Yellow Adder's tongue,* as given in Gray's Manual, is much better.

Ornithogalum umbellatum, *L.*
Danvers, 1852 (J. L. Russell); "rare at the Bowler farm, Dr. J. M. Nye" (Tracy); Rockport (C. W. Pool); Amesbury (Miss Perley). Escaped into meadows. (Nat. from Eu.)

Allium Canadense, *Kalm.* (WILD GARLIC.)
Orne's Point, Salem; "occasional in Lynn" (Tracy); Georgetown (Mrs. Horner), Andover, etc. Frequent on moist banks.

Muscari botryoides, *Mill.* (GRAPE HYACINTH.)
Essex County, 1817 (Wm. Oakes). The only locality known for this plant is in Danvers, where it grows quite plentifully in company with Arisæma triphyllum and Trillium cernuum. (Adv. from Eu.)

Hemerocallis fulva, *L.*
The common red lily of old gardens. This has escaped into fields and by the roadside, in many places often forming large patches.

Hemerocallis flava, *L.* (COMMON YELLOW GARDEN LILY.)
Also escaped in a few places. Both species of Hemerocallis were introduced from Europe.

JUNCACEÆ.

(RUSH FAMILY.)

Luzula pilosa, *Willd.*
Essex county (herb. P. A. S., Oakes). The station is not known.

Luzula campestris, *DC.*
Very common in dry fields.

Juncus effusus, *L.* (SOFT RUSH.)
Very common in wet places.

Var. conglomeratus. (Gray's Manual.)
Salem Great Pastures, Ipswich (Oakes), etc. Frequent.

Juncus Balticus, *Dethard.*
Essex county, 1825 (Oakes in Proc. E. I., Vol. 1, p. 271), Byfield, (J. H. Sears), Kernwood, Newbury, etc. Near salt water.

Juncus marginatus, *Rostkovius.*
Ipswich (Oakes), Lawrence, West Gloucester, etc. Very common in moist soil.

Juncus bufonius, *L.*
Common in most parts of the county. This is the most widely distributed of all the Junci, being found in all parts of the world.

Juncus Gerardi, *Loisel.* (BLACK GRASS.)
Everywhere along the coast. Extensively cut, and much used under the name of "marsh hay;" being the most valuable product of our salt marshes.

Juncus tenuis, *Willd.*
Common and very variable. Found in all situations.

Juncus Greenii, *Oakes & Tuckerman.*
Ipswich (Tuckerman in Hov. Mag., Vol. IX, 1843), Salem, Newbury, Salisbury, etc.; Ipswich (herb. P. A. S., Oakes). Frequent in dry and sandy places, and even on salt marshes.

Juncus pelocarpus, *E. Meyer.*
West Gloucester, Salem Great Pastures, Wenham, etc. Frequent.

Juncus militaris, *Bigelow.* (BAYONET RUSH.)
Abundant around Chebacco pond; Crane pond, West Newbury;

Hood's pond; Topsfield, etc. Scarce in many places. "Discovered (1823) by Mr. Benj. D. Greene, growing plentifully in a pond at Tewksbury" (Bigelow's Fl. Bost., 2d ed., 1824). This was the locality from which the species was originally described by Dr. Bigelow.

Juncus acuminatus, *Michx.*, var. **legitimus.** (Gray's Manual.)
Lanesville, West Newbury, etc. Often found bearing proliferous heads. The typical form is not found here.

Juncus articulatus, *L.*
West Newbury, Salem Great Pastures, and also at Wenham (W. P. Conant). Found on wet banks.

Juncus Canadensis, *J. Gay*, var. **longicaudatus.** (Gray's Manual.)
A variable species. Moist places. Common.

Var. **coarctatus.** (Gray's Manual.)
In situations similar to the last. Common.

PONTEDERIACEÆ.

(PICKEREL-WEED FAMILY.)

Pontederia cordata, *L.* (PICKEREL-WEED.)
Ponds and streams. Common.

Schollera graminea, *Willd.*
Pleasant and Wenham ponds, Wenham, and a few other localities, but not yet noticed in flower.

XYRIDACEÆ.

(YELLOW-EYED-GRASS FAMILY.)

Xyris flexuosa, *Muhl.*, *Chapm.*
Shores of ponds. Not rare.

Var. **pusilla.** (Gray's Manual.)
Crooked pond, Boxford; Chebacco pond. Not so abundant as the last.

ERIOCAULONACEÆ.

(PIPEWORT FAMILY.)

Eriocaulon septangulare, *Withering.* (PIPEWORT.)
Common on the shores of ponds, and often in water two feet deep, the stems lengthening to accommodate themselves to the increased depth.

CYPERACEÆ.
(Sedge Family.)

Note.—The following list of sedges and also that of the grasses will be found quite large and perhaps nearly complete. These very useful and interesting plants are too often neglected by the student, and what is most singular, the people of the country towns are seldom familiar with any considerable number of species. Frequently, persons are met with who do not distinguish between the two orders, not knowing which are sedges and which are grasses. The writer is indebted to Mr. Woodbury P. Conant, of Wenham, and Mr. Chas. E. Faxon, of Boston, for valuable assistance in determining the more difficult species, particularly those of the genus Carex.

Cyperus diandrus, *Torr.*
Shores of rivers and ponds. Common.

Cyperus Nuttallii, *Torr.*
Sandy places, near salt water.

Cyperus Engelmanni, *Steud.*
Borders of Fresh pond, Cambridge (C. E. Faxon).

Cyperus erythrorhizos, *Muhl.*
Near the city of Lawrence. Abundant along the shore of the Merrimac, above the dam. First noticed (J. R.), Sept. 26, 1877. Perhaps introduced with cotton brought to the mills at Lowell, or above, upon the same river. Common at the south.

Cyperus dentatus, *Torr.*
Plum Island, West Gloucester, Lawrence, and along the shores of the Merrimac. Frequent.

Cyperus phymatodes, *Muhl.*
Frequent. Natural to low grounds. There is a serious danger that this plant may become the pest here, which its cogener, Cyperus rotundus, has become at the south, where several plantations have been abandoned to it. C. phymatodes will readily establish itself in cultivated fields; it has been particularly noticed at Danvers and West Newbury. At the latter town, Mr. W. P. Conant finds that, when once established, the plant will form, about ten inches below the surface of the ground, a mat of roots, underground stems, and tubers, which, being below the usual ploughing, remain undisturbed throughout the season, several yards and fields being already overrun by this very persistent perennial sedge. Unless some immediate and radical means are resorted to, this plant is likely to rival the White weed, or the more recently introduced Cone flower, as a farm nuisance.

Cyperus strigosus, *L.*
Moist places. Common.

Cyperus Michauxianus, *Schultes.*
This species is frequently found near Boston, and undoubtedly grows within the county limits.

Cyperus Grayii, *Torr.*
Plum Island (Proc. E. I., Vol. I, p. 273); same station, 1878 (W. P. Conant).

Cyperus filiculmis, *Vahl.*
Dry sandy soil. Common.

Cyperus fuscus, *L.*
Near Revere beach, 1877 (Herbert A. Young). This species is also found in New Jersey. (Adv. from So. Eu.)

Dulichium spathaceum, *Pers.*
Swamps and borders of ponds. Common.

Eleocharis Robbinsii, *Oakes.*
Sluice pond, Lynn (C. E. Faxon). Probably not uncommon.

Eleocharis tuberculosa, *R. Br.*
Manchester (Oakes); Danvers (J. H. Sears). Rare.

Eleocharis obtusa, *Schultes.*
Wet places and borders of ponds.

Eleocharis olivacea, *Torr.*
Long pond, Saugus (Herbert A. Young).

Eleocharis palustris, *R. Br.*
Wenham and Chebacco ponds, Crane pond, West Newbury, etc.

Eleocharis tenuis, *Schultes.*
Meadows, ponds, etc. Very common and variable.

Eleocharis acicularis, *R. Br.*
Wet places. Common.

Eleocharis pygmæa, *Torr.*
Salem Great Pastures, near Atlantic Car Works; near Coffin's beach, West Gloucester. Scarce.

Eleocharis Engelmanni, *Steudel.*
Sluice pond, Lynn, Sept. 1879 (Herbert A. Young).

Var. detonsa, *Gray.*
Found at Winter pond, Winchester, 1878 (E. H. Hitchings).

Scirpus planifolius, *Muhl.*
Danvers, 1878 (J. H. Sears).

Scirpus subterminalis, *Torr.*
Crane pond, West Newbury; Chebacco pond; "Wenham, Pleasant pond, 1823" (memo. Dr. Chas. Pickering).

Scirpus pungens, *Vahl.*
Found near both fresh and salt water. Common.

Scirpus Torreyi, *Olney.*
Lawrence, along the shores of the Merrimac, above the dam.

Scirpus validus, *Vahl.* (GREAT BULRUSH.)
Newbury (Oakes); Pleasant pond, Wenham, growing seven or eight feet high; West Gloucester, Byfield, etc. This species is sometimes found in brackish waters.

Scirpus debilis, *Pursh.*
Lawrence, above the dam on the Merrimac (J. R.). Scarce.

Scirpus supinus, *L.*, var. **Hallii.** (Gray's Manual).
Winter pond, Winchester, 1878 (E. H. Hitchings). Beyond the county limits, yet within the latitude, and but a few miles to the westward.

Scirpus maritimus, *L.* (SEA CLUB-RUSH.)
Common in brackish water. Specimens were collected at Kernwood, Salem, 1879, five feet high, bearing panicles three times the ordinary size.

Scirpus fluviatilis, *Gray.*
Ipswich river, Aug., 1876 (J. R.). Rare.

Scirpus sylvaticus, *L.*
"Along brooks, eastern Massachusetts, W. Boott," etc. (Gray's Manual, 5th ed.), Lawrence, August 5, 1879. Specimens are frequently met with where the styles are two-cleft, and, hence, the achenium lenticular.

Scirpus microcarpus, *Presl.*
Shore of the Merrimac, from Lawrence to Newburyport. Not very common.

Scirpus atrovirens, *Muhl.*
Shores of Ipswich and Merrimac rivers, North Andover, Danvers (J. H. Sears). Scarce.

Scirpus Eriophorum, *Michx.*
Meadows. Very common and variable. Flowering from June to September.

Scirpus lineatus, *Michx.*
Essex county, about 1824 (memo. Dr. Chas. Pickering). Not in the county herbarium P. A. S.

Eriophorum alpinum, *L.*
Hamilton (Oakes); Boxford (J. R.); Georgetown (Mrs. C. N. S. Horner). Abundant where it grows at all.

Eriophorum vaginatum, *L.*
Long pond, Saugus (Herbert A. Young). It is quite probable that this species may be found in other similar localities.

Eriophorum Virginicum, *L.*
Bogs. Frequent.

Var. **album** (Gray's Manual).
In similar situations. Common.

Eriophorum polystachyon, *L.* (COMMON COTTON-GRASS.)
Common in meadows. Variable.

Eriophorum gracile, *Koch.*
Topsfield (Oakes); Boxford; Cedar pond, Wenham; Beaver pond, Beverly; Danvers (J. H. Sears). Not very common.

Fimbristylis autumnalis, *Rœm. & Schultes.*
Boxford (J. R.); Danvers, 1878 (J. H. Sears); Salem Great Pastures (W. P. Conant).

Fimbristylis capillaris, *Gray.*
Common in dry open places.

Rhynchospora fusca, *Rœm. & Schultes.*
Manchester (Oakes); Beaver pond, Beverly (W. P. Conant); Middleton pond (J. R.).

Rhynchospora alba, *Vahl.*
Meadows. Frequent.

Rhynchospora glomerata, *Vahl.*
Wet places, borders of ponds, etc. Very common.

Cladium mariscoides, *Torr.*
Ipswich (Oakes); Chebacco pond (J. R.); Hood's pond, Topsfield (J. H. Sears). Scarce.

Carex polytrichoides, *Muhl.*
West Newbury and Wenham (W. P. Conant).

Carex bromoides, *Schk.*
Ipswich (Oakes); West Newbury (W. P. Conant).

Carex siccata, *Dew.*
Danvers (J. H. Sears); Salem Great Pastures.

Carex teretiuscula, *Good.*
Danvers (J. H. Sears); Beaver pond, Beverly.

Carex vulpinoidea, *Michx.*
Meadows. Common.

Carex stipata, *Muhl.*
Salem, Danvers (J. H. Sears), Merrimac, etc. Common.

Carex sparganioides, *Muhl.*
Essex Co. (Oakes).

Carex muricata, *L.*
Salem Great Pastures, Kernwood (W. P. Conant), etc.

Carex cephalophora, *Muhl.*
Boxford, 1878; West Newbury (W. P. Conant).

Carex Muhlenbergii, *Schk.*
Boxford, Wenham, Beverly, etc. Quite common.

Carex rosea, *Schk.*
Meadows, etc. Common.

Carex retroflexa, *Muhl.*
Boxford, June, 1878 (C. E. Faxon).
Carex trisperma, *Dew.*
Topsfield (J. R.).
Carex canescens, *L.* (in part).
Meadows. Common.
Carex exilis, *Dew.*
First "noticed in marshes around Lake Wenham; the specimens were taken by Oakes to Dewey who named the species" (Dr. Chas. Pickering, Chron. Hist. Pl., p. 1063). Hamilton, Danvers (Oakes), Newbury (W. P. Conant), brooks of Saugus and Lynnfield (C. E. Faxon), Beverly, Boxford, etc. In peat bogs. Not very common.
Carex sterilis, *Willd.*, and
Carex stellulata, *Good.*, var. **scirpoides** (Gray's Manual).
These species are both found in the county, but great difficulty is experienced in separating them.
Carex scoparia, *Schk.*
Meadows. Common. Variable.
Carex lagopodioides, *Schk.*
Moist places in shade. Frequent.
Carex cristata, *Schw.*, var. **mirabilis,** *Boott.*
West Newbury (W. P. Conant), Danvers (J. H. Sears).
Carex adusta, *Boott.* (Gray's Manual, 5th ed.).
West Newbury (W. P. Conant).
Carex silicea, *Olney.* C. fœnea, *Willd.*, var. sabulonum (Gray's Manual.)
Sandy places along the coast. Common.
Carex straminea, *Schk.*
Widely distributed and very variable.
Var. **typica** (Gray's Manual).
Occasionally in fields, etc.
Var. **tenera** (Gray's Manual).
East Haverhill (Rock's Village), (W. P. Conant).
Var. **aperta** (Gray's Manual).
Quite a distinct form which may prove worthy of being raised to the rank of a species. Perfecting earlier than the other varieties. Danvers (J. H. Sears), Ipswich river banks, etc.
Carex umbellata, *Schk.*
Abundant near Boston, and undoubtedly in Essex county (C. E. Faxon).
Carex alata, *Torr.*
West Newbury (W. P. Conant); Andover, near Haggett's pond.
Carex vulgaris, *Fries.*
Low ground. Common.

Carex stricta, *Lam.*
Common. The sedge so commonly forming the hummocks in meadows.

Carex maritima, *Vahl.*
"Salt marsh at Orne's point, Salem (memo. Dr. Chas. Pickering); Newburyport (memo. Prof. Brainard).

Carex salina, *Wahl.*
Near Newburyport (memo. Prof. Brainard). Both this species and the last are found in many places near the coast of Massachusetts.

Carex crinita, *Lam.*
Low ground. Common.

Carex Buxbaumii, *Wahl.*
Saugus river, near Lynnfield (Herbert A. Young); Ipswich (Oakes); Burley farm meadow, Danvers, Andover. Scarce.

Carex aurea, *Nutt.*
Edge of a meadow, Salem Great Pastures, June 17, 1879 (J. R.).

Carex livida, *Willd.*
Swamp at Boxford. Rare.

Carex panicea, *L.*
Danvers, 1879 (J. H. Sears).

Carex pallescens, *L.*
Meadows. Common.

Carex conoidea, *Schk.*
West Newbury (W. P. Conant); (Salem (J. R.); Danvers (J. H. Sears).

Carex gracillima, *Schw.*
Ipswich (Oakes), Wenham, etc. Frequent.

Carex virescens, *Muhl.*
Copses, etc. Frequent.

Carex retrocurva, *Dew.*
West Newbury, 1879 (W. P. Conant).

Carex digitalis, *Willd.*
West Newbury, 1879 (W. P. Conant).

Carex laxiflora, *Lam.*
Partly shady places. Common and very variable.

Var. **intermedia,** *Boott.*
Salem, West Newbury (W. P. Conant).

Var. **stylofiexa,** *Boott.*
Groveland, etc. (W. P. Conant).

Var. **blanda** (Gray's Manual).
West Newbury (W. P. Conant). The most common form.

Var. **latifolia,** *Boott.*
Chebacco woods, etc. Frequent.

Carex Emmonsii, *Dew.*
Ipswich (Oakes).

Carex Pennsylvanica, *Lam.*
Hills and copses. Common. One of our earliest spring flowers.

Carex varia, *Muhl.*
West Newbury (W. P. Conant).

Carex præcox, *Jacq.*
"Salem and Ipswich, Mass." (Gray's Manual); Salem (Oakes); Orne's point, Salem, 1824 (Dr. Chas. Pickering); Dedham (C. E. Faxon); Salem Great Pastures, in several places; Swampscott, near the "Willows." This species is considered by Gray to have been introduced from Europe. When first noticed by Dr. Chas. Pickering (1824), it "seemed to be growing wild."

Carex scabrata, *Schw.*
Near Haggett's pond, Andover, June 27, 1879; Danvers (J. H. Sears).

Carex debilis, *Michx.*
Meadows. Common.

Carex flava, *L.*
Meadows, etc. Common.

Carex Œderi, *Ehrh.*
"By European colonists carried to northeast America." "Only in grass-grown clearings in the environs of Salem" (Dr. Chas. Pickering, Chron. Hist. Pl. p. 1024).

This is farther explained in a letter, where he says "C. Œderi (determined by English Botany), I used to find in the pasture land of my grandfather's farm (Wenham) clearly introduced."

There seems to be some uncertainty regarding this species, as the form here found does not perfectly correspond to the description. Rocks at Pigeon Cove, 1877; Salem Great Pastures in wet land (J. R.); Danvers (J. H. Sears), Magnolia. Considered by Prof. Gray to be an American plant.

Carex filiformis, *L.*
Peat meadows. Common.

Carex lanuginosa, *Michx.*
Swamps. Frequent.

Carex vestita, *Willd.*
Moist, sandy soil. Frequent.

Carex riparia, *Curtis.*
Burley woods, Danvers (J. H. Sears); West Newbury (W. P. Conant).

Carex comosa, *Boott.*
Swampscott, Wenham (W. P. Conant), Danvers (J. H. Sears). Meadows.

Carex hystricina, *Willd.*
Near Crane pond, West Newbury, Aug. 1879.

Carex tentaculata, *Muhl.*
Meadows, etc. Very common.

Carex intumescens, *Rudge.*
Danvers (J. H. Sears), West Newbury (W. P. Conant), etc. Frequent.

Carex lupulina, *Muhl.*
Swampy places. Frequent.

Carex folliculata, *L.*
Wet and shady places. Quite common.

Carex squarrosa, *L.*
One specimen near Sutton's Mills station, North Andover, Aug. 5, 1879 (J. R.).

Carex utriculata, *Boott.*
Essex County (Oakes); Beaver pond, Beverly; West Newbury (W. P. Conant). Frequent.

Carex bullata, *Schk.*
Near Haggett's pond, Andover; West Newbury (W. P. Conant); Danvers (J. H. Sears).

Carex monile, *Tuckerman.*
Burley woods, Danvers, 1879 (J. H. Sears); Saugus river, June 18, 1879 (C. E. Faxon); West Newbury (W. P. Conant).

Carex oligosperma, *Michx.*
Essex County (Oakes, in herb. P. A. S.). This specimen may have been collected outside the limits of the county.

GRAMINEÆ.

(GRASS FAMILY.)

Leersia Virginica, *Willd.* (WHITE GRASS.)
Oak Island, Chelsea (Bigelow's Fl. Bost., 2d ed., 1824); near Sutton's Mills station, North Andover. Abundant at the last named place.

Leersia oryzoides, *Swartz.* (RICE CUT-GRASS.)
Borders of brooks, etc. Common.

Zizania aquatica, *L.* (INDIAN RICE.)
Outlet of Pleasant pond (Dr. Chas. Pickering), outlet of Wenham pond (W. P. Conant), Ipswich river from Topsfield to Ipswich, Merrimac river. Abundant. An exceedingly beautiful plant when in flower.

Alopecurus pratensis, *L.* (MEADOW FOXTAIL.)
Everywhere. (Nat. from Eu.)
Alopecurus geniculatus, *L.* (FLOATING FOXTAIL.)
West Newbury (Conant), Ipswich (Oakes), Danvers (J. H. Sears). (Nat. from Eu.)
Alopecurus aristulatus, *Michx.* (WILD FOXTAIL.)
West Newbury, June 19, 1878 (W. P. Conant).

Phleum pratense, *L.* (TIMOTHY.)
Everywhere. Introduced from Europe. Dr. Chas. Pickering considered this as likely to prove an American grass, as one of foreign origin.

Vilfa aspera, *Beauv.*
Ipswich (Oakes), Danvers (J. H. Sears).
Vilfa vaginæflora, *Torr.*
Poor soil. Frequent.

Sporobolus cryptandrus, *Gray.*
Ipswich (Oakes), Nahant.
Sporobolus serotinus, *Gray.*
Moist places. Common. It is probable that in future botanies, this and the last species will be found under Vilfa.

Agrostis perennans, *Tuckerman.* (THIN-GRASS.)
Danvers (J. H. Sears), Oak Island (W. P. Conant).
Agrostis scabra, *Willd.* (HAIR GRASS.)
Dry places. Common.
Agrostis canina, *L.* (BROWN BENT-GRASS.)
Wenham (W. P. Conant).
Agrostis Spica-venti, *L.*
In a field at West Newbury (W. P. Conant). (Adv. from Eu.)
Agrostis vulgaris, *With.* (RED-TOP.)
Everywhere. Considered an American grass. It has also been introduced from Europe. This species is called Herd's-grass in Pennsylvania, the name given in New England to Phleum pratense.
Agrostis alba, *L.* (WHITE BENT-GRASS.)
"Meadows and fields, a valuable grass: naturalized from Europe, also indigenous on river banks, N. New York and northward" (Gray's Manual.) A variety of this species is met with in the county which may prove to have another place, as it appears quite distinct.

Polypogon Monspeliensis, *Desf.* (BEARD GRASS.)
Collected at Hampton Beach by Oakes and Robbins. This European grass should be looked for at Salisbury, as the above locality is but a few miles north. (Nat. from Eu.)

Cinna arundinacea, *L.,* var. **pendula,** *Gray.*
Lawrence, Danvers (J. H. Sears); West Newbury (W. P. Conant) etc. Not very common. Forms are occasionally met with which are quite near the typical plant.

Muhlenbergia sobolifera, *Trin.*
Beverly, 1878 (J. H. Sears). Rare.

Muhlenbergia glomerata, *Trin.*
Boxford, West Newbury, Ballardvale (J. H. Sears). Swamps.

Muhlenbergia Mexicana, *Trin.*
Ballardvale (J. H. Sears); Salem (Jona. Tucker); Wenham (J. R.).

Muhlenbergia sylvática, *Torr. & Gray.*
West Newbury (W. P. Conant); Beverly (J. H. Sears). Wenham.

Muhlenbergia Willdenovii, *Trin.*
Burley woods, Danvers, Sept. 1877.

Brachyelytrum aristatum, *Beauv.*
Bank of Shawsheen, Andover, near the meeting-house (J. R.); West Newbury (W. P. Conant).

Calamagrostis Canadensis, *Beauv.* (BLUE JOINT-GRASS.)
Swamps and river banks. Common.

Calamagrostis Pickeringii, *Gray.*
Collected at Andover, near Haggett's pond, June 27, 1879 (J. R.), and found abundantly in the same locality, 1880. First observed by Mr. H. Little with Dr. Pickering about 1824, at the White Mountains, in N. H.

Calamagrostis Nuttalliana, *Steud.*
Ipswich (Oakes), Wenham swamp, West Gloucester, etc. Not very common.

Calamagrostis arenaria, *Roth.*
Beaches. Common. This is the species which is so useful in holding in check the shifting sands of our coast. Large sums have been expended in planting it on Cape Cod by the state government.

Oryzopsis melanocarpa, *Muhl.*
Chebacco woods, Georgetown. Scarce.

Oryzopsis asperifolia, *Michx.*
Manchester (Oakes), Chebacco woods, May 17, 1875.

Oryzopsis Canadensis, *Torr.*
Manchester (Oakes); Boxford (several stations); Middleton (J. H. Sears).

Stipa avenacea, *L.* (BLACK OAT GRASS; FEATHER GRASS.)
"Dry wooded hills, near the Andover turnpike, Medford" (Bigelow's Fl. Bost., 2d ed., 1824); Blue Hills, Milton (Chas. E. Faxon). This ought to be found in the southwestern portion of the county.

Aristida dichotoma, *Michx.* (POVERTY GRASS.)
Frequent in poor soil.
Aristida gracilis, *Ell.*
Danvers, 1819 (Oakes in Hov. Mag., Vol. XIII), and in the same region, 1878 (J. H. Sears).
Aristida purpurascens, *Poir.*
On W. P. Upham's farm, Peabody (W. P. Conant).
Aristida tuberculosa, *Nutt.*
Plum Island, 1829 (Oakes), and same station, near a farm house on the Ipswich end of the Island, 1876 (J. R.).

Spartina cynosuroides, *Willd.*
Ipswich river banks, and near the salt water everywhere.
Spartina juncea, *Willd.*
Salt marshes along the entire shore. Common.
Spartina stricta, *Roth.* (SALT MARSH GRASS.)
Between tides in all the inlets. This is the "Thatch" of the farmers.

Bouteloua oligostachya, *Torr.*
Near the old carpet factory, Tapleyville, 1880 (J. H. S. and J. R.). Probably introduced from the west with wool.

Eleusine Indica, *Gærtn.* (DOG'S-TAIL GRASS.)
This is very common at New York city and Philadelphia, but never noticed here until 1878, when it seemed to have established itself on Pennsylvania pier, Salem. (Probably Nat. from India.)

Tricuspis purpurea, *Gray.* (SAND GRASS.)
Turkey Hill, Ipswich, Newbury; Nahant beach.

Dactylis glomerata, *L.* (ORCHARD GRASS.)
Fields and roadsides. Common. (Nat. from Eu.)

Eatonia Pennsylvanica, *Gray.*
Burnt land "Blind Hole," Danvers (J. H. Sears); West Newbury (W. P. Conant).

Glyceria Canadensis, *Trin.* (RATTLESNAKE GRASS.)
Common in open swamps.
Glyceria obtusa, *Trin.*
Swamps and wet places. Frequent.
Glyceria elongata, *Trin.*
Essex county (Oakes).
Glyceria nervata, *Trin.* (FOWL MEADOW GRASS.)
Common in meadows.
Glyceria pallida, *Trin.*
In shallow water. Boxford, etc. Not very common.

Glyceria aquatica, *Smith.* (REED MEADOW GRASS.)
"Ipswich river marsh" (Dr. Chas. Pickering); West Newbury (W. P. Conant); Lynn (J. H. Sears).

Glyceria fluitans, *R. Br.*
Shallow water. Various parts of the county, but not common.

Glyceria acutiflora, *Torr.*
In similar situations to the last. Not common.

Glyceria maritima, *Wahl.*
Along the coast. Common.

Glyceria distans, *Wahl.*
Plum Island (Oakes).

Brizopyrum spicatum, *Hook.*
Frequent on the salt marshes.

Poa annua, *L.* (LOW SPEAR-GRASS.)
Very common.

Poa compressa, *L.* (WIRE-GRASS.)
Common in dry places.

Poa serotina, *Ehrhart.* (FALSE REDTOP.)
Sometimes called Fowl Meadow-grass. Essex county (Oakes), Wenham, Byfield, etc.

Poa pratensis, *L.* (COMMON MEADOW-GRASS; JUNE GRASS; KENTUCKY BLUE-GRASS.)
Common, both in a cultivated and wild state.

Poa trivialis, *L.* (ROUGHISH MEADOW-GRASS.)
Ipswich (Oakes); West Newbury (W. P. Conant); environs of Salem (Dr. Chas. Pickering, Chron. Hist. Pl., p. 279). (Nat. from Eu.)

Eragrostis poæoides, *Beauv.*, var. **megastachya.** (Gray's Manual).
First noticed on the "dump," So. Boston, by Mr. C. E. Faxon; Marblehead, 1878 (Mr. Bartlett). (Nat. from Eu.)

Eragrostis capillaris, *Nees.*
"Possibly introduced into New England by the aboriginal tribes; observed by myself in the environs of Salem, chiefly in cultivated ground" (Dr. Chas. Pickering, Chron. Hist. Pl., p. 810). Probably noticed about 1824.

Eragrostis tenuis, *Gray.*
"In the same situations with the preceding, in the environs of Salem, and as far as Philadelphia" (Dr. Chas. Pickering, Chron. Hist. Pl., p. 810). Probably noticed at the same date. There is no other record of this and the last species in Essex county.

Eragrostis pilosa, *Beauv.*
Salem, two stations (W. P. Conant). (Nat. from Eu.)

Eragrostis pectinacea, *Gray*, var. **spectabilis.** (Gray's Manual).
Topsfield, Newburyport, Ballardvale (J. H. Sears). Scarce.

Briza media, *L.* (QUAKING GRASS.)
Danvers (Oakes); Peabody, Boxford, Salem Great Pastures. Scarce. (Adv. from Eu.)

Festuca tenella, *Willd.*
Ship Rock, Peabody (memo. Rev. J. L. Russell); Pigeon Cove; East Haverhill (Rock's Village). Scarce.

Festuca ovina, *L.* (SHEEP'S FESCUE.) Var. **duriuscula.**
The most common form.
Var. **rubra** (Gray's Manual).
Plum Island (Oakes).

Festuca elatior, *L.* (MEADOW FESCUE.)
Very common and quite variable. (Nat. from Eu.)

Festuca nutans, *Willd.*
Boxford, Danvers (J. H. Sears.) Scarce.

Bromus secalinus, *L.* (CHEAT; CHESS.)
Beverly, Danvers, Ipswich, etc. Frequent. Var. (awnless), Ipswich (Oakes). (Adv. from Eu.)

Bromus mollis, *L.* (SOFT CHESS.)
Among other species in a grass-plat in Salem. (Adv. from Eu.)

Bromus racemosus, *L.* (UPRIGHT CHESS.)
Essex County (Oakes), West Gloucester, Salem, etc. Scarce. (Adv. from Eu.)

Bromus ciliatus, *L.*
Ipswich (Oakes), Wenham, Salem, Beverly, etc. More abundant than the last.

Phragmites communis, *Trin.*
In the vicinity of Topsfield. Probably introduced in this locality.

Lolium perenne, *L.* (COMMON DARNEL.)
West Newbury (W. P. Conant); Salem, etc. Not very common. (Nat. from Eu.)

Lolium temulentum, *L.*
"In old barley fields and gathered with it" (Oakes, Hovey's Mag., Vol. VII). "Grain noxious; almost the only instance of the kind among grasses" (Gray's Manual.) (Adv. from Eu.)

Triticum vulgare, *L.* (WHEAT.)
Found growing on wharves and railroad track-beds near Salem and Newburyport. (Escaped.)

Triticum repens, *L.* (COUCH-GRASS; QUICK-GRASS.)
Everywhere common and variable.

Var. intermedium, *Fries.* (nearly) *Gray.*
West Gloucester. The spikes of this variety are much larger every way than the common form.

Secale cereale, *Willd.* (RYE.)
Derby Wharf, Salem. (Escaped.)

Hordeum vulgare, *L.* (BARLEY.)
Salem, Essex, etc. Waste places. (Escaped.)

Hordeum jubatum, *L.* (SQUIRREL-TAIL GRASS.)
Marblehead Neck, Beverly, Ipswich (Oakes), etc. Frequent near the shore.

Elymus Virginicus, *L.*
Frequent. Usually nearer the shore than the next.

Elymus Canadensis, *L.*
Merrimac river banks, from Lawrence to Newburyport.

Elymus striatus, *Willd.*
Ipswich (Oakes), West Gloucester, Georgetown (Mrs. C. N. S. Horner), along the Merrimac river banks, etc. Frequent.

Gymnostichum Hystrix, *Schreb.* (BOTTLE-BRUSH GRASS.)
Swampscott, Hamilton, Oak Island, Revere. Scarce.

Danthonia spicata, *Beauv.*
Dry soil. Common.

Danthonia compressa, *Austin.*
Frequent near Boston (C. E. Faxon), and undoubtedly an Essex county species. It has but recently been separated from the last.

Avena sativa, *L.* (COMMON OAT.)
Waste places, railroad beds, wharves, etc. (Escaped.)

Avena striata, *Michx.*
"Manchester" (S. P. Fowler), banks of the Merrimac, West Newbury. Rare.

Aira flexuosa, *L.* (HAIR-GRASS.)
Dry soil. Frequent.

Arrhenatherum avenaceum, *Beauv.*
"Near Salem" (Dr. Chas. Pickering), probably about 1824; old fields, Danvers (J. H. Sears), and sometimes cultivated. Scarce. (Nat. from Eu.)

Holcus lanatus, *L.* (VELVET GRASS.)
"Sparingly naturalized in the environs of Salem" (Dr. Chas. Pickering, Chron. Hist. Pl., p. 655). This was probably about 1824. It is now abundant, spreading fast in moist soil, and, unless care is taken, will injure much valuable grass land. (Nat. from Eu.)

Hierochloa borealis, *Rœm. & Schultes.* (VANILLA GRASS.)
Salisbury, Byfield (Mrs. Horner), Beverly, Salem near Pickman farm, etc. Scarce. Most frequently found near the shore.

Anthoxanthum odoratum, *L.* (SWEET VERNAL-GRASS.)
Fields. Common. (Nat. from Eu.)

Phalaris Canariensis, *L.* (CANARY-GRASS.)
Streets and rubbish heaps near the larger towns. Frequent. (Adv. from Eu.)

Phalaris arundinacea, *L.*
Low ground. Frequent.

Var. **picta.** (STRIPED-GRASS; RIBBON-GRASS.)
Common in old gardens. A large patch of this variety escaped from cultivation on the bed of the E. R. R., at Beverly, shows every variation from the strongest marked striping to the typical green form.

Paspalum setaceum, *Michx.*
Fields. Common.

Cynosurus cristatus, *Willd.*
Among other species in a grass-plat, Salem. "The stems of this grass are used for the manufacture of Leghorn hats." It has been cultivated to some extent in the county (Flint's Grasses). (Adv. from Eu.)

Panicum Germanicum (*Flint's Grasses.*) (HUNGARIAN MILLET.)
Cultivated, and found near fields, although not to be considered by any means as established, Danvers (J. H. Sears). This seems to be the Setaria Germanica, *Beauv.*, of some writers, yet very much larger than the ordinary form, and may be only a large variety of Setaria Italica. (Int. from Eu.)

Panicum miliaceum, *Willd.* (MILLET.)
First noticed by Mr. C. E. Faxon, in 1876, on the "dump" at So. Boston; near West Beach, Beverly, 1879 (W. P. Conant).

Panicum glabrum, *Gaudin.*
Streets and yards. Common. (Nat. from Eu.)

Panicum sanguinale, *L.* (COMMON CRAB GRASS; FINGER-GRASS.)
Similar places. Common. (Nat. from Eu.)

Panicum agrostoides, *Spreng.*
Meadows, and the shores of rivers and ponds. Common.

Panicum proliferum, *Lam.*
Near the B. R. B. & L. R. R. station, Lynn, and West Newbury (W. P. Conant), are the only localities yet noticed for this species in the county. In each case it appears as if introduced.

Panicum capillare, *L.* (OLD WITCH GRASS.)
Fields, etc. Common.
Panicum virgatum, *L.*
Beaches among rocks, and inland in wet sandy places. Frequent.
Panicum latifolium, *L.*
Shady places. Lanesville, Danvers (Oakes), Groveland, etc.
Panicum clandestinum, *L.*
Banks of Merrimac and Ipswich rivers. Not rare.
Panicum dichotomum, *L.*
Very common and variable. We have three distinct forms, some of which, according to Gray, may prove to be distinct species, and it is quite probable that P. pauciflorum, *Ell.*, may have been included here.
(*a*). Rigid, large-flowered.
(*b*). Less rigid, taller, small-flowered.
(*c*). Autumnal, prostrate.
All are common.
Panicum depauperatum, *Muhl.*
Dry hilly places. Common.
Panicum Crus-galli, *L.* (BARNYARD GRASS.)
Waste places and cultivated fields. Common. (Nat. from Eu.)
Var. **hispidum** (Gray's Manual).
Similar situations. Common. (Nat. from Eu.)

Setaria verticillata, *Beauv.*
A street grass in Salem, etc. (Adv. from Eu.)
Setaria glauca, *Beauv.*
Fields and waste places. Common. (Adv. from Eu.)
Setaria viridis, *Beauv.*
Fields and streets. Common. (Adv. from Eu.)
Setaria Italica, *Kunth.*
Sometimes called Millet. Cultivated lands. (Nat. from Eu.)

Andropogon furcatus, *Muhl.*
Dry places in sandy soil. Frequent.
Andropogon scoparius, *Michx.*
Dry soil. Common.

Sorghum nutans, *Gray.* (INDIAN GRASS.)
Ipswich (Oakes), Danvers (J. H. Sears). Scarce.

Zea Mays, *L.* (INDIAN CORN.)
A reduced form is often found on wharves and railroad beds, having the flowers developed when only a foot high.

VASCULAR CRYPTOGAMS.

LYCOPODIACEÆ.

(CLUB-MOSS FAMILY.)

Lycopodium lucidulum, *Michx.*
Essex woods, Rockport (Frank Lufkin), Peabody, Andover, Georgetown, etc. Frequent.

Lycopodium inundatum, *L.*
Near Coffin's beach, West Gloucester; Plum Island; Beaver pond, Beverly.

Var. Bigelovii, *Tuckerman.*
Chebacco pond. Both forms are quite common.

Lycopodium annotinum, *L.*
Chebacco and Essex woods. Abundant where it is found at all. The above locality was probably known to Oakes fifty years ago.

Lycopodium dendroideum, *Michx.* (GROUND PINE.)
In woods. Frequent.

Var. obscurum (Gray's Manual).
With the other form. Common.

Lycopodium clavatum, *L.*
Frequent in various parts of the county.

Lycopodium complanatum, *L.*
Commonly called "Evergreen." In most woods. Formerly very common, but so largely gathered for decorations that it is becoming scarce within a radius of five or six miles of our larger cities.

Selaginella rupestris, *Spring.*
Rocky hills. Frequent.

Selaginella apus, *Spring.*
Boxford, Danvers, Topsfield. Not very common.

Isoetes echinospora, *Durieu,* var. **muricata** (Gray's Manual). (QUILLWORT.)
Beaver pond, Beverly; Chebacco; Crane pond, West Newbury; Pleasant pond, Wenham. Rather scarce. Other species should be looked for.

RHIZOCARPEÆ.
(WATER-FERN FAMILY.)

Marsilia quadrifolia, *L.* (MARSILIA.)
First discovered at Bantam lake, Conn., some years ago. It has become established in some ponds near Boston, and for awhile in one of the ponds near Legge's hill, Salem. It is feared, however, that on account of a severe drought a year or two since, it has been destroyed.

OPHIOGLOSSACEÆ.
(ADDER'S-TONGUE FAMILY.)

Ophioglossum vulgatum, *L.*
Moist, not boggy places, in grass, near the edges of swamps, Boxford (two stations near Crooked pond, J. R. and J. H. S.); Beverly, 1872 (J. H. Emerton, etc.); Danvers, 1879 (J. H. Sears). Rather scarce.

Botrychium lanceolatum, *Angs.*
Georgetown, Aug., 1875 (Mrs. C. N. S. Horner); West Newbury, 1879 (W. P. Conant); Magnolia, 1880 (C. J. Sprague); Middleton, 1880 (J. R.).

Botrychium matricariæfolium, *A. Br.*
West Newbury, 1879 (W. P. Conant); Middleton (J. H. S. and J. R.), the small form often confounded with B. simplex; Boxford, 1880, the larger form.

Botrychium Virginianum, *Sw.* (RATTLESNAKE FERN.)
Beaver pond woods, Beverly, 1870 (J. R.); West Newbury (W. P. Conant), Danvers, Haverhill, Georgetown, Boxford, etc. Not very common.

Botrychium ternatum, *Sw.*
Moist pastures and even in dry soil. A most variable species.

(*a*). Var. **rutæfolium,** *Eaton.*
This is usually called var. lunarioides, which is a southern form and not found here.

(*b*). Var. **australe,** *Eaton.*
This is the most highly developed form; and specimens, very nearly resembling the fine plants often sent from California, have been collected near Danvers by J. H. Sears, and others hardly less fine, at Georgetown by Mrs. C. N. S. Horner.

(*c*). Var. **obliquum,** *Milde.*
The most common form with quite sharply pointed segments.

(*d*). Var. **dissectum,** *Milde.*
Nearly as common as the last. The frond is cut and divided in a very beautiful manner.

EQUISETACEÆ.

(Horsetail Family.)

Equisetum arvense, *L.* (Common Horsetail.)
Moist places. Very common. On the railroad beds, a prostrate, much branched form occurs, caused by the continual passing of trains, which prevents the plant attaining its ordinary height.

Equisetum sylvaticum, *L.*
In shady places. Frequent. A very graceful plant.

Equisetum limosum, *L.*
Open wet places. Frequent.

Equisetum hyemale, *L.* (Scouring Rush.)
Near Cold Spring, Salem (Dr. G. A. Perkins); Methuen, on the bed of the M. & L. R. R.; Danvers, on the bed of the Newburyport R. R.; "moist woods, Lynn" (Bigelow's Fl. Bost., 1814). The least common species.

FILICES.

(Ferns.)

Note.—This order and the Ophioglossaceæ have been rearranged to conform to the "Systematic Fern List," but very recently (Oct., 1880) published by Professor Daniel C. Eaton of New Haven.

Polypodium vulgare, *L.* (Polypody.)
Covering shaded rocks. Common.

Var. Cambricum (English authors).
A specimen corresponding to this variety was found some years since, at Andover, by Mr. Jackson T. Dawson.

Adiantum pedatum, *L.* (Maidenhair.)
Towns of the Merrimac valley, frequent; and in small stations elsewhere.

Pteris aquilina, *L.* (Common Brake.)
In half shade on well drained soil. Quite common. A variety having the fronds several times pinnate, but not ternate, is occasionally met with; the same form is found near Boston (E. H. Hitchings).

Woodwardia Virginica, *Smith.* (Virginian Chain-Fern.)
Borders of ponds, extending its rhizomes under the water; swamps, etc. Frequent.

Woodwardia angustifolia, *Smith.* (Narrow-leaved Chain-Fern.)
Magnolia swamp, Gloucester (J. H. S. and J. R.); "Essex county" (memo. Dr. Chas. Pickering). Rare.

Asplenium Trichomanes, *L.*
Clefts of rocks. Common.
Asplenium ebeneum, *Ait.*
Dry, rocky places, usually in pine woods. Common and somewhat variable.
Asplenium thelypteroides, *Michx.*
Ipswich (Oakes), Lynnfield (Rev. J. L. Russell), Swampscott (J. R.). Scarce.
Asplenium Filix-fœmina, *Bernh.* (LADY FERN.)
Very variable.
(*a*). Three to four feet high, twelve to sixteen inches wide, lower pinnæ smallest. Moist shady places. Common.
(*b*). Var. **Rhæticum,** *Moore.* One to two feet high, four to six inches wide, rigid. Banks and roadsides. Common.
(*c*). Var. **Michauxii,** *Mett.* Two to three feet high, stem red, foliage delicate. Essex woods. Scarce.
(*d*). Two to three feet high, lower pinnæ largest. West Newbury, Beverly. Scarce. Various other intermediate forms are found, sometimes mimicking Aspidium spinulosum, and again resembling Aspidium Noveboracense.

Phegopteris polypodioides, *Fée.* (BEECH FERN.)
Danvers, Beverly, Essex, Manchester, Middleton, Haverhill, West Newbury. Scarce, and never found in large quantities.
Phegopteris hexagonoptera, *Fée.*
Georgetown (Mrs. C. N. S. Horner), Chebacco woods (J. H. S. and J. R.). Rare.
Phegopteris Dryopteris, *Fée.*
Haverhill (W. P. Conant), Boxford, Middleton, Beverly, Essex, Georgetown (Mrs. C. N. S. Horner). Scarce, the stations being small.

Aspidium Thelypteris, *Sw.*
Moist places. Common.
Aspidium Noveboracense, *Sw.* (NEW YORK FERN.)
Damp places in the woods. Common.
Aspidium spinulosum, *Sw.*
A common and very variable evergreen fern.
(*a*). Var. **vulgare,** *Eaton.* (The typical form.)
Essex, Georgetown, Beverly, Haverhill, etc. Frequent.
(*b*). Var. **dilatatum,** *Eaton.*
This seems to be a strongly developed form of the typical plant, the lower pinnæ being very broad. This variety does not grow in the county to the perfection attained farther north, yet the form here found is sufficiently near it to be included.

(c). **Var. intermedium,** *Eaton.* Aspidium Americanum, *Davenport* (Am. Nat.).
This is the most common form, and seems quite distinct from either of the above varieties. Frequent in moist woods. Other varieties may be found corresponding very well with the numerous named forms, in which English fern books abound, most of which are not constant and are unworthy of a separate name.

Aspidium Boottii, *Tuckerman.* A. spinulosum, *Sw.*, var. Boottii (Gray's Manual).
Swamps. Frequent. A very distinct species.

Aspidium cristatum, *Sw.*
Situations similar to the last. Frequent.

Var. Clintonianum, *Eaton.*
Wenham swamp (J. H. S. and J. R.). Scarce.

Aspidium marginale, *Sw.*
Rocky places in the shade. Common.

Aspidium acrostichoides, *Sw.* (CHRISTMAS FERN.)
Rocky woods. Common.

Var. incisum (Gray's Manual).
West Newbury (W. P. Conant), Georgetown, Beverly. Occasional.

Cystopteris fragilis, *Bernh.*
Damp rocks. Common. Often disappearing by the latter part of August.

Onoclea Struthiopteris, *L.* Struthiopteris Germanica, *Willd.* (Gray's Manual.) (OSTRICH FERN.)
North Andover, near Sutton's Mills station and the Merrimac river (abundant), Georgetown, Boxford. Scarce.

Onoclea sensibilis, *L.* (SENSITIVE FERN.)
Moist places. Very common.

Var. obtusilobata (Gray's Manual, etc.).
Frequently met with. This is merely a fertile frond which is partly sterile and is common to all species, having contracted fruiting fronds or portions of fronds, as in O. Struthiopteris, Osmunda (var. frondosa), Lygodium, and in Botrychium (Ophioglossaceæ). It is an abnormal condition and not a true variety.

Woodsia obtusa, *Torr.*
Rocky places in half shade. Rather scarce.

Woodsia Ilvensis, *R. Br.*
Rocky hills, in exposed situations; particularly abundant in the vicinity of Salem.

Dicksonia pilosiuscula, *Willd.* D. punctilobula, *Kunze* (Gray's Manual, 5th ed.). (HAY-SCENTED FERN.)
Shady places in moist woods. Common.

Lygodium palmatum, *Swartz.* (CLIMBING FERN.)
Lynn, near Saugus; said to have been first detected in this locality by Mr. B. F. Johnson (herb. P. A. S., Mr. G. E. Emery).

Osmunda regalis, *L.* (ROYAL FERN; FLOWERING FERN.)
Swamps. Common.

Osmunda Claytoniana, *L.* (INTERRUPTED FLOWERING FERN.)
Swamps and moist places. Common, more especially so in Topsfield and North Andover.

Osmunda cinnamomea, *L.*
Moist pastures. Common. The abnormal form (var. frondosa) is frequently found among the fruiting fronds, and should be looked for in June.

MUSCINEÆ.

The following list of the *Musci* and *Hepaticeæ* is made up from the herbarium of the Peabody Academy of Science, chiefly collected by the writer in 1877. Unfortunately, there are few persons in this vicinity interested in the lower orders of plants, hence the data from which to make up a general county list are meagre. It is not possible that this enumeration should be nearly complete, as many species, perhaps even common ones, may not have been collected, but it very well represents the general character of the Muscineæ of the region, and will be found to contain many rare and interesting species. The collection was examined, and the species determined by Mr. Coe F. Austin of New Jersey, well known as an authority upon these plants, and whose unexpected death, during the past season, is universally regretted by botanists.

MUSCI.

(Mosses.)

Suborder SPHAGNACEÆ.

Sphagnum acutifolium, *Ehrh.*
Chebacco (in fruit, May 3, 1877).
Var. purpureum.
Boxford (Arthur R. Stone).
Var. intermedium.
Danvers (J. H. Sears).
Sphagnum cuspidatum, *Ehrh.*
Middleton, Peabody, North Andover, Amesbury. Submerged.
Var. Torreyanum.
Middleton, North Andover.
Var. recurvum.
Magnolia swamp, Gloucester.
Sphagnum subsecundum, *Nees.*
Boxford.
Sphagnum palustre, *Ehrh.* (S. cymbifolium, *Dill.*)
Magnolia swamp, Gloucester. A very large species.
Var. squarrulosum.
Swampscott.

Sphagnum papillosum, *Lindb.*
Essex woods and Middleton. "Probably not sufficiently distinct from S. palustre" (Austin).

Sphagnum laricinum, *Spruce,* var. **platyphyllum** (*Sulliv.*) *Austin.* (S. neglectum, var. Aust. Musc. Appal.), (S. platyphyllum, Sullivant in a letter to Austin, 1868).
In a pool, on the line between Boxford and North Andover. A curious submerged species resembling an alga.

Suborder BRYACEÆ.

Tribe WEISIEÆ.

Weisia viridula, *Brid.*
Essex; Beverly, June, 1877 (Mary K. Robinson); Amesbury, etc. Frequent.

Tribe DICRANEÆ.

Dicranum scoparium, *L.*
Rocky woods. Common.

Dicranum Schraderi, *Web. & Mohr.*
Boxford.

Dicranum congestum, *Brid.*
Middleton.

Dicranum flagellare, *Hedw.*
Boxford.

Dicranum montanum, *Hedw.*
Essex, Boxford.

Dicranum fulvum, *Hk.*
Swampscott, Essex.

Ceratodon purpureus, *Brid.*
Rocky pastures and roadsides. Common.

Tribe LEUCOBRYEÆ.

Leucobryum vulgare. (L. glaucum, *Hampe*).
In woods. Frequent.

Tribe FISSIDENTEÆ.

Fissidens osmundioides, *Hedw.*
Boxford.

Fissidens taxifolius, *Hedw.*
Swampscott (E. M. Shepard). This specimen varies from the typical form.

Fissidens adiantoides, *Hedw.*
Boxford.

Tribe TRICHOSTOMEÆ.

Trichostomum tortile, *Schrad.*
Near "Ship Rock," Peabody.

Trichostomum pallidum, *Hedw.*
"Prospect Hill," Peabody.

Barbula ruralis, *Hedw.*
"On rocks, Nahant, Mass., D. Murray" (Gray's Manual with mosses).

Barbula papillosa, *Wils.*
"Trunks of elm trees, Mass., Rev. J. L. Russell" (Gray's Manual with mosses).

Tribe POTTIEÆ.

Pottia truncata, *Br. & Sch.*
Garden walks, Salem (Rev. J. L. Russell). In greenhouses.

Tribe TETRAPHIDEÆ.

Tetraphis pellucida, *Hedw.*
Old stumps, and at the bases of trees. Common.

Tribe ORTHOTRICHEÆ.

Orthotrichum anomalum, *Hedw.*
"Rocks, near Salem, Mass., Lesquereux" (Gray's Manual with mosses). Swampscott.

Orthotrichum Hutchinsiæ, *Smith.*
Marblehead; "Rubley hill," Beverly.

Orthotrichum Lescurii, *Austin.*
Swampscott.

Orthotrichum crispulum, *Hornsch.*
Near Salem (Rev. E. C. Bolles).

Tribe GRIMMIEÆ.

Schistidium apocarpum, *Br. & Sch.* (Grimmia apocarpa.)
Nahant (B. D. Greene).

Schistidium maritimum, *Br. & Sch.* (Grimmia maritima.)
Nahant (B. D. Greene).

Tribe HEDWIGIEÆ.

Hedwigia ciliata, *Ehrh.*
On bowlders and rocky places. Common.

Tribe BUXBAUMIEÆ.

Diphyscium foliosum, *Web. & Mohr.*
Hill, near Chebacco pond; Danvers (J. H. Sears). Rare.

Tribe POLYTRICHEÆ.

Atrichum angustatum, *Beauv.*
Moist places. Common.

Polytrichum commune, *L.*
Moist places. Very common.

Polytrichum piliferum, *Schreb.*
Salem, etc. Frequent.

Polytrichum juniperinum, *Hedw.*
Salem Neck.

Tribe BRYEÆ.

Aulacomnion palustre, *Schwægr.*
Moist banks. Frequent.

Var. **rupestre** (memo. C. F. Austin).
Swampscott.

Aulacomnion androgynum, *Schwægr.*
Prospect hill, Peabody, the form with pseudopodia only. Rare.

Aulacomnion heterostichum, *Br. & Sch.*
Swampscott, Chebacco.

Bryum nutans, *Schreb.*
Wenham, Middleton (Rev. E. C. Bolles).

Bryum pseudo-triquetrum, *Schwægr.*
Boxford.

Bryum bimum, *Schreb.*
Middleton.

Bryum intermedium, *Brid.*
A specimen collected in Swampscott was marked by Mr. Austin, as a variety of this species.

Bryum cæspiticium, *L.*
Common on moist soil.

Mnium affine, *Bland.*
Boxford.

Mnium hornum, *Hedw.*
Essex county (Rev. E. C. Bolles).

Mnium punctatum, *Hedw.*
Danvers, Boxford.

Mnium cuspidatum, *Hedw.*
Peabody, Boxford, Danvers, Beverly.

Tribe BARTRAMIEÆ.

Bartramia pomiformis, *Hedw.*
Moist banks and rocks. Frequent.
Bartramia fontana, *Brid.*
Wet places. Occasional.

Tribe FUNARIEÆ.

Funaria hygrometrica, *Hedw.*
Garden paths, burnt ground, etc. Very common.
Physcomitrium pyriforme, *Br. & Sch.*
Danvers, Salem (in greenhouses).

Tribe FONTINALEÆ.

Fontinalis antipyretica, *L.*, var. **gigantea.**
Georgetown (Mrs. C. N. S. Horner).
Fontinalis Lescurii, *Sulliv.*
Boxford, Topsfield, Salem Great Pastures.
Fontinalis Novæ-Angliæ, *Sulliv.* (in Gray's Manual, 3d ed.).
Boxford. Mr. Austin marks this specimen as a little doubtful. All the species of this genus are found growing upon stones, wood, or some other substance, in flowing water. Rockport (Prof. Thos. P. James).

Dichelyma capillaceum, *Bryol. Europ.*
Brook at Beverly Farms.

Anomodon rostratus, *Br. & Sch.*
At the bases of trees, Boxford, Swampscott, etc.
Anomodon attenuatus, *Hub.*
Boxford, Swampscott.

Leskea polycarpa, *Hedw.*
Bases of trees; Wenham, 1876 (Mary K. and Lucy P. Robinson), Beverly (Rev. E. C. Bolles).
Leskea obscura, *Hedw.*
Ipswich (Oakes); Salem (Rev. J. L. Russell).
Leskea rostrata, *Hedw.*
Swampscott (Rev. E. C. Bolles).

Tribe THELIEÆ.

Thelia hirtella (*Hedw.*), *Sulliv.*
At the bases of trees. Frequent.
Thelia asprella (*Schimp.*) *Sulliv.*
In similar places to the last. Frequent.

Thelia Lescurii, *Sulliv.*
Boxford.

Tribe PYLAISÆEÆ.

Pylaisæa intricata, *Bryol. Europ.*
On trees, Chebacco woods.

Tribe CYLINDROTHECIEÆ.

Cylindrothecium cladorrhizans, *Bryol. Europ.*
Beverly woods.

Tribe NECKEREÆ.

Neckera pennata, *Hedw.*
Gloucester, Beverly Farms.

Tribe CLIMACIEÆ.

Climacium Americanum, *Brid.*
Moist woods. Frequent.

Tribe HYPNEÆ.

Hypnum delicatulum, *L.* ("H. tamariscinum of many authors," Austin).
A very graceful species growing over rocks and dead trees, in moist shady places. Common.

Hypnum recognitum, *Hedw.* ("H. delicatulum of authors," Austin.)
Boxford; Magnolia station, Gloucester; Chebacco.

Hypnum paludosum, *Sulliv.*
Near Cedar pond, Wenham, 1849 (Rev. J. L. Russell); Boxford.

Hypnum brevirostre, *Ehrh.*
Swampscott.

Hypnum triquetrum, *L.*
Chebacco, May 17, 1875 (fine fruit), rocky moist banks. Frequent.

Hypnum splendens, *Hedw.*
Beverly woods, May 11, 1876 (fine fruit). Moist places in old woods. Not rare.

Hypnum strigosum, *Hoffm.*
Middleton.

Hypnum rusciforme, *Weis.*
Chebacco woods.

Hypnum Schreberi, *Willd.*
Borders of woods, Chebacco.

Hypnum cordifolium, *Hedw.*
Danvers, 1852 (Rev. J. L. Russell), Wenham swamp.

Hypnum Crista-Castrensis, *L.*
Lynn, 1856 (Rev. J. L. Russell), Essex, Amesbury, etc.

Hypnum cupressiforme, *L.*
Swampscott (Rev. E. C. Bolles), Beverly, Wenham, Chebacco (fine fruit), etc. Frequent.

Var. **resupinatum** (memo. C. F. Austin).
Beverly Farms woods (very fine), pendent over rocks, in fruit.

Hypnum imponens, *Hedw.*
Chebacco woods.

Hypnum reptile, *Michx.*
Common lane, Beverly.

Hypnum curvifolium, *Hedw.*
Boxford; Peabody, near "Ship Rock." Frequent.

Hypnum Haldanianum, *Grev.*
Near "Ship Rock," Peabody; Beverly, Chebacco.

Hypnum lætum, *Brid.*
Swampscott.

Hypnum acuminatum, *Beauv.*
Peabody, Beverly.

Var. **setosum** (Gray's Manual with mosses).
Danvers.

Hypnum rutabulum, *L.*
Beverly, Swampscott.

Hypnum plumosum, *L.*
"Rubley hill," on the Chebacco road in Wenham; near Coffin's beach, West Gloucester, growing in sand.

Hypnum populeum, *Hedw.* (H. plumosum, var. populeum, *C. F. Austin*).
Beverly Farms.

Hypnum velutinum, *L.*
Beverly, near "Pride's Crossing."

Hypnum Novæ-Angliæ, *Sulliv. & Lesqx.*
Boxford.

Hypnum stellatum, *Schreb.*
Salem, 1852 (Rev. J. L. Russell).

Hypnum hispidulum, *Brid.*
Middleton.

Hypnum serpens, *Hedw.*
Danvers, Salem. Variable.

Hypnum adnatum, *Hedw.*
Peabody.

Hypnum denticulatum, *L.*, var. **lætum** (*C. F. Austin*).
Swampscott, at the base of a tree.

Hypnum sylvaticum, *L.*
Near "Magnolia swamp," Gloucester.

Hypnum striatellum, *Brid.*
  Middleton. Marked as a little doubtful.
Hypnum campestre (memo. C. F. Austin).
  Swampscott (Rev. E. C. Bolles).
Hypnum chrysophyllum, *Brid.*
  Swampscott.

Plagiothecium turfaceum, *Lindb.*
  Beverly.

HEPATICÆ.

(LIVERWORT FAMILY.)

Tribe RICCIACEÆ.

Riccia natans, *L.*
  Floating on the water of a spring, "Blind Hole," Danvers, Dec., 1878 (J. H. Sears).
Var. terrestris (memo. C. F. Austin).
  On the east bank of Foster's pond, near North Beverly depot.
Riccia fluitans, *L.*
  Floating in a brook, Wenham.
Var. terrestris (memo. C. F. Austin).
  Banks of the fresh pond, near Coffin's beach, West Gloucester.
Riccia lutescens, *Schwein.*
  Wet places in Manchester woods (J. H. Sears).

Tribe MARCHANTIACEÆ.

Marchantia polymorpha, *L.*
  Common in shady wet places.

Lunularia vulgaris, *Michx.*
  Greenhouse beds, and rarely in wet gardens, Salem. (Adv. from Eu.)

Preissia commutata, *Nees.*
  Wet rocks, E. R. R. cut near Magnolia station, Gloucester.

Conocephalus conicus (*L.*) *Dumort.* (Fegatella conica, *Corda.*)
  Marblehead (S. B. Buttrick); Boxford.

Fimbriaria tenella, *Nees.*
  Wet soil, with Selaginella apus, at the base of "Bald hill," Middleton.

Tribe JUNGERMANNIACEÆ.

Pellia epiphylla, *Nees.*
  Wet banks. Marblehead (E. M. Shepard), Chebacco, Beverly.

Geocalyx graveolens, *Nees.*
Old log, Boxford woods.

Chiloscyphus polyanthos, *Corda.*, var. **rivularis** (memo. C. F. Austin).
On a log, in a brook, Boxford.

Lophocolea heterophylla, *Nees.*
With mosses and lichens on trees, Boxford.

Harpanthus scutatus, *Spruce.* (Jungermannia scutata, *Weber.*)
Beverly woods.

Jungermannia Schraderi, *Martius.*
Middleton (Rev. E. C. Bolles).

Jungermannia acuta (memo. C. F. Austin).
Magnolia swamp, Gloucester. This specimen is a little doubtful.

Jungermannia ventricosa, *Dicks.*
Beverly Farms woods.

Scapania nemorosa, *Nees.*
Moist slopes and rocks in the woods. Frequent.

Scapania undulata, *Nees & Montagne.*
Magnolia swamp, Gloucester. This specimen varies from the typical form.

Frullania Grayana, *Montagne.*
On rocks and tree trunks. Common.

Frullania Eboracensis, *Lehm.*
Bark of trees, Boxford.

Madotheca platyphylla, *Dumortier.*
Moist, rocky places in shade. Common.
Var. ――
Near "Ship Rock," Peabody.

Radula complanata, *Dumortier.*
Beverly Farms woods; Boxford, May 29, 1877 (fine fruit).

Blepharozia ciliaris (*L.*), *Dumortier.* (Ptilidium ciliare, *Nees.*)
Near "Ship Rock," Peabody (very fine); bases of trees and rocks, fruiting abundantly in April and May. Common.
Var. **teretiuscula** (memo. C. F. Austin).
Beverly woods. Rare.

Trichocolea tomentella, *Nees.*
Beverly Farms woods. The exact locality is a little uncertain.

Mastigobryum trilobatum, *Nees.*
Moist woods. Frequent.

CHARACEÆ.

Plants growing under water in ponds and slow streams, resembling Algæ, with which they are classed by some authors. Some of the species have the very disagreeable odor of sulphuretted hydrogen. The classification is taken from a recent paper upon the subject by Byron D. Halsted, D. S. (Proc. Bost. Nat. Hist., Vol. XX, p. 169; March 5, 1879). The specimens have all been identified by Prof. Farlow and Dr. Halsted, to whom the writer would here acknowledge his indebtedness.

NITELLEÆ.

Nitella opaca, *Ag.*
Specimens of a Nitella collected at Swampscott in a spring hole, May 11, 1879, are referred to this species by Dr. Farlow, who says: —"The specimens are all male, and the plant seems to be diœcious. If I am correct in supposing it to be diœcious, the species belongs to the group of N. capitata, to which N. syncarpa, N. capitata, and N. opaca belong. The color is lighter than most forms of N. opaca, but I am disposed to refer it to that species which it resembles in habit" (Letter, Nov. 16, 1879). This determination has since been confirmed by Nordstedt (note, Dr. Farlow, Jan. 8, 1880).

Nitella flexilis, *Ag.*
Common in ponds and streams. Chebacco pond; Merrimac river, above the dam at Lawrence; in a somewhat brackish stream near Burley Farm, Danvers; Wenham pond. Dr. Halsted refers to certain Essex county specimens as approaching the

Var. nidifica. (Chara glomerulifolia, *A. Br.*), and (N. flexilis, var. subcapitata, *A. Br.*).
He also speaks of specimens of this species as having been collected in the "Merrimac (Green)," probably B. D. Greene.

Nitella gracilis (*Sw.*), *Ag.*
Chebacco pond, near Whipple's boat-landing. The specimens are all of the delicate form.

Nitella (Tolypella) intricata (*Roth.*), *Ag.*
In a brook below an old mill in Boxford, Aug. 13, 1876. "The Tewksbury specimens resemble those gathered by Mr. Wright in the Leom river, Texas, being in size between the small specimens just mentioned (from Louisiana), and the large showy plants of Boxford,[1] Mass. (Robinson)" (Halsted's Char.).

CHAREÆ.

Chara coronata, var. **Schweinitzii**, *A. Br.*
Wenham pond, Aug., 1875; Chebacco pond. The typical form of this species grows near Boston.

Chara gymnopus, var. **elegans**, *A. Br.*
Pleasant pond, Wenham, Aug., 1875. "The variety elegans seems to have been first found in America by Oakes. In the Gray herbarium is a specimen collected by him, which, however, bears neither date nor locality. It probably came from Essex county, Mass., where the variety has been recently re-discovered by Mr. John Robinson, who reports that it is rather common in Chebacco lake.[2] Among the specimens in alcohol sent by Mr. Robinson, a deviation from the variety form was found. The difference is confined entirely to the leaves, the more slender nature of which, as well as its much shorter bracts at the sterile joints, is noticeable. The excessive length of the bracts around the sexual organs is in contrast with those in the ordinary form. The most striking peculiarity of the plant is its not bearing both sexual organs at the same joint. Usually, a single sporangium and antheridium were found on each leaf, sometimes two sporangia and no antheridium, or two autheridia and no sporangium, but never more than two of these organs on a leaf, and these at the second and third joints from the base. This arrangement of the sexual organs may be considered as steps towards a diœcious species, first, in the separation of the organs to different joints, and then to different leaves. This form is of great interest, and would furnish a fine subject to one who loves such deviations, and can procure an abundance of specimens" (Halsted's Char.).

[1] Since the above was written, Dr. T. F. Allen of New York, whose monograph of the North American Characeæ is now in process of publication, has requested specimens of the Boxford plant, and from those sent, he considers the species to be
Nitella polyglochin, *A. Br.*,
an East Indian species which, as he says, supports a theory regarding certain Characeæ, that he intends presenting before the Torrey Botanical Club very soon.

[2] This should read Pleasant pond, Wenham.

Chara fragilis, *Desv.* (Chara fœtida of authors.)
Pleasant pond, Wenham; Salem, in a brook on the Turnpike (Mr. H. F. King) many years since. This Chara possesses the most disagreeable odor of any found in this region.

Var. delicatula, *A. Br.*
Wenham, Aug. 3, 1876.

Var. ———
Collected in Beaver pond, Beverly, July 17, 1879. Dr. Farlow in a letter of Nov. 16, 1879, says, of this plant: — " I hesitate between C. fragilis and C. tenuispora. I think it a variety of the former, but uncomfortably near the latter. I have compared it with European forms of C. tenuispora and the specimens are a good deal smaller than the European forms of that species. It is, however, more spiny than C. fragilis in which the spines are merely rudimentary in most cases. The stipular crown has the upper whorl more developed than in C. fragilis and, in this respect, the species also approaches C. tenuispora. In spite of the differences named I should consider it a form of C. fragilis, judging by the general habit."

Chara sejuncta, *A. Br.*
Among other specimens from Essex county, collected in 1879. Determined by Nordstedt through Dr. W. G. Farlow.

THALLOPHYTES.

LICHENS.

The writer, owing to his want of knowledge of this order, had felt compelled very reluctantly to omit the Lichens from the flora, as no authentically-named collection existed of sufficient size to make a list of any value. But in the present year Mr. Charles J. Sprague most kindly offered to take what specimens were already in the herbarium, name them, and, adding those species he had himself collected in the vicinity of Magnolia, prepare the whole list.

For this kindness the writer desires to express his obligations, feeling it a great pleasure to be able to present the list from the pen of so competent an authority as Mr. Sprague.

(The following list is, probably, but a portion of the species occurring in Essex county, as the region has not been thoroughly explored. It comprises only such as are positively known to grow there.)

1. Ramalina, *Ach.*, *De Not.*

1. calicaris, *Fr.*
    var. fastigata, *Fr.*
    var. farinacea, *Fr.*
2. pollinaria, *Ach.*

2. Cetraria, *Ach.*, *Fr.*

1. Islandica, *Ach.*
2. ciliaris, *Ach.*
3. lacunosa, *Ach.*
    var. Atlantica, *Tuck.*
4. aleurites (*Ach.*), *Th. Fr.*
5. Oakesiana, *Tuck.*
6. juniperina (*L.*), *Ach.*
    var. Pinastri, *Ach.*

3. Evernia, *Ach.*, *Mann.*

1. furfuracea (*L.*), *Mann.*
2. prunastri (*L.*), *Ach.*

4. Usnea, *Ach.*

1. barbata (*L.*), *Fr.*
    var. florida, *Fr.*
    var. hirta, *Fr.*
    var. rubiginea, *Mx.*
2. trichodea, *Ach.*

5. Alectoria (*Ach.*), *Nyl.*

jubata (*L.*), *Fr.*
    var. chalybeiformis, *Ach.*

6. Theloschistes, *Norm.*, *Tuck.*

1. chrysopthalmus (*L.*), *Norm.*
2. parietinus (*L.*), *Norm.*
    var. lychneus, *Schær.*
3. concolor (*Dicks.*).

7. Parmelia (*Ach.*), *De Not.*

1. perforata (*Jacq.*), *Ach.*
    var. crinita, *Tuck.*
2. tiliacea (*Hoffm.*), *Flœrk.*
3. Borreri, *Turn.*
    var. rudecta, *Tuck.*
4. saxatilis (*L.*), *Fr.*
5. pertusa (*Schrank.*), *Schær.*
6. physodes (*L.*), *Ach.*
7. colpodes, *Ach.*
8. caperata (*L.*), *Ach.*
9. conspersa (*Ehr.*), *Ach.*
10. olivacea (*L.*), *Ach.*

8. Physcia (*Fr*). *Th. Fr.*

1. aquila (*Ach.*), *Nyl.*
    var. detonsa, *Tuck.*
2. pulverulenta (*Schreb.*), *Nyl.*
3. speciosa (*Wulf.*, *Fr.*)
    var. hypoleuca, *Ach.*
4. stellaris (*L.*), *Nyl.*
    var. tribacia, *Fr.*
    var. hispida, *Fr.*
5. cæsia (*Hoffm.*), *Nyl.*
6. obscura (*Ehrh.*), *Nyl.*
    var. ciliata, *Schær.*
    var. adglutinata (*Flk.*), *Nyl.*

9. Pyxine, *Fr.*

cocoes (*Sw.*), *Nyl.*
   var. sorediata, *Tuck.*

10. Umbilicaria, *Hoffm.*

1. Pennsylvanica, *Hoffm.*
2. pustulata (*L.*), *Hoffm.*
3. Dillenii, *Tuck.*
4. Muhlenbergii (*Ach.*), *Tuck.*

11. Sticta (*Schreb.*), *Delis.*

1. crocata (*L.*), *Ach.*
2. quercizans (*Mx.*), *Ach.*
3. pulmonaria (*L.*), *Ach.*
4. scrobiculata, *Ach.*
5. aurata, *Ach.*
6. glomerulifera (*Lightf.*), *Delis.*

12. Nephroma, *Ach.*

1. lævigatum, *Ach.*
2. tomentosum (*Hoffm.*), *Kbr.*

13. Peltigera (*Hoffm.*), *Fée.*

1. apthosa (*L.*), *Hoffm.*
2. canina (*L.*), *Hoffm.*
3. polydactyla (*Neck.*), *Hoffm.*
4. rufescens (*Neck.*), *Hoffm.*
5. horizontalis (*L.*), *Hoffm.*

14. Pannaria, *Delis.*

1. rubiginosa (*Thunb.*), *Ach.*
   var. conoplea, *Fr.*
2. tryptophylla (*Ach.*), *Mass.*
3. microphylla (*Sw.*), *Del.*
4. leucosticta, *Tuck.*
5. brunnea (*Sw.*), *Mass.*
6. molybdæa (*Pers.*), *Tuck.*
   var. cronia, *Nyl.*

15. Ephebe, *Fr.*

pubescens (*Ach.*), *Fr.*

16. Lichina, *Ag.*, *Mont.*

confinis (*Mull.*), *Ag.*(?)

17. Pyrenopsis, *Nyl.*

phæococca, *Tuck.*

18. Collema (*Hoffm.*), *Fr.*

1. flaccidum, *Ach.*
2. nigrescens (*Huds.*), *Ach.*

19. Leptogium, *Fr.*

1. pulchellum (*Ach.*), *Nyl.*
2. tremelloides, *Fr.*
3. chloromelum (*Sw.*), *Nyl.*
4. myochroum (*Ehrh.*), *Schær.*
    var. saturninum (*Dicks.*), *Tuck.*

20. Placodium (*D C.*), *Næg. & Hepp.*

1. elegans (*Link.*), *D C.*
2. murorum (*Hoffm.*), *D C.*
3. vitellinum (*Ehrh.*), *Hepp.*
4. cerinum (*Hedw.*), *Næg.*
5. microphyllinum, *Tuck.*, *ined.*
6. aurantiacum (*Lightf.*), *Næg.*
7. ferrugineum (*Huds.*), *Hepp.*
    var. nigricans, *Nyl.*

21. Lecanora, *Ach.*, *Tuck.*

1. rubina (*Vill.*), *Ach.*
2. muralis (*Schreb.*), *Schær.*
3. pallescens (*L.*), *Fr.*
4. tartarea (*L.*), *Ach.*
5. subfusca (*L.*), *Ach.*
6. pallida (*Schreb.*), *Schær.*
7. varia (*Ehrh.*), *Fr.*
8. ventosa, *Ach.*
9. elatina, *Ach.*
    var. ochrophæa, *Tuck.*
10. cinerea (*L.*), *Fr.*
11. cervina (*Pers.*), *Sommerf.*
    var. discreta, *Fr.*

22. Rinodina, *Mass., Stizenb.*

1. oreina (*L.*), *Mass.*
2. sophodes (*Ach.*).
    var. confragosa, *Nyl.*
    var. exigua, *Nyl.*
3. constans (*Nyl.*), *Tuck.*

23. Pertusaria, *D C.*

1. pertusa (*L.*), *Ach.*
2. leioplaca (*Ach.*), *Schær.*
3. velata (*Turn.*), *Nyl.*
4. pustulata (*Ach.*), *Nyl.*

24. Conotrema, *Tuck.*

urceolatum (*Ach.*), *Tuck.*

25. Gyalecta (*Ach.*), *Anzi.*

1. lutea (*Dicks.*), *Tuck.*
2. pineti (*Schrad.*), *Tuck.*

26. Urceolaria (*Ach.*), *Flot.*

scruposa (*L.*), *Ach.*

27. Stereocaulon, *Schreb.*

1. paschale, *Laur.*
2. condensatum, *Laur.*

28. Cladonia, *Hoffm.*

1. pyxidata (*L.*), *Fr.*
2. fimbriata (*L.*), *Fr.*
    var. radiata, *Fr.*
3. gracilis (*L.*), *Fr.*
    var. verticillata, *Fr.*
    var. elongata, *Fr.*
4. turgida (*Ehrh.*), *Hoffm.*
5. furcata (*Huds.*), *Fr.*
    var. racemosa, *Flk.*
    var. subulata, *Flk.*
6. cenotea, *Ach.*
    var. furcellata, *Fr.*
7. squamosa, *Hoffm.*
    var. cæspiticia, *Nyl.*
8. rangiferina (*L.*), *Hoffm.*
    var. sylvatica, *L.*
    var. alpestris, *L.*

9. uncialis (*L.*), *Fr.*
10. Boryi, *Tuck.*
11. cornucopioides (*L.*), *Fr.*
12. cristatella, *Tuck.*

29. Bæomyces, *Pers.*, *Nyl.*

1. roseus, *Pers.*
2. byssoides (*L.*), *Schœr.*

30. Biatora, *Fr.*

1. granulosa (*Ehrh.*), *Pœtsch.*
2. viridescens (*Schrad.*), *Fr.*
3. vernalis (*L.*), *Th. Fr.*
4. sanguineo-atra (*Fr.*), *Tuck.*
5. exigua (*Chaub.*), *Fr.*
6. Nylanderi, *Anzi.*
7. uliginosa (*Schrad.*), *Fr.*
8. rivulosa (*Ach.*), *Fr.*
9. lucida (*Ach.*), *Fr.*
10. milliaria (*Fr.*), *Tuck.*
11. rubella (*Ehrh.*), *Rabenh.*
12. umbrina (*Ach.*), *Tuck.*
13. chlorantha, *Tuck.*
14. resinæ, *Fr.* It is doubtful if this be a lichen.

31. Heterothecium, *Flot.*, *Tuck.*

sanguinarium (*L.*), *Flot.*

32. Lecidea (*Ach.*), *Fr.*

1. albo-cœrulescens, *Fr.*
2. contigua, *Fr.*, *Nyl.*
3. enteroleuca, *Ach.*
4. melancheima, *Tuck.*

33. Buellia, *DeNot.*, *Tuck.*

1. parasema (*Ach.*), *Kbr.*
2. dialyta (*Nyl.*).
3. myriocarpa (*D C.*), *Mudd.*
4. turgescens (*Nyl.*).
5. vernicoma, *Tuck.*
6. petræa (*Flot.*), *Tuck.*

34. Opegrapha (*Humb.*), *Ach.*, *Nyl.*

1. varia (*Pers.*), *Fr.*
2. vulgata, *Ach.*, *Nyl.*
3. viridis, *Pers.*

35. Xylographa, *Fr.*, *Nyl.*
opegraphella, *Nyl.*

36. Graphis, *Ach.*, *Nyl.*
1. scripta (*L.*), *Ach.*
2. dendritica, *Ach.*

37. Arthonia, *Ach.*, *Nyl.*
1. pyrrhula, *Nyl.*
2. lecideella, *Nyl.*
3. astroidea, *Ach.*, *Nyl.*
4. punctiformis, *Ach.*

38. Mycoporum (*Flot.*), *Nyl.*
pycnocarpum, *Nyl.*

39. Acolium (*Fée*), *DeNot.*
tigillare (*Ach.*), *DeNot.*

40. Calicium, *Pers.*, *Fr.*
microcephalum (*Sw.*), *Turn. & Borr.*

41. Endocarpon, *Hedw.*, *Fr.*
miniatum (*L.*), *Schær.*

42. Trypethelium, *Spreng.*, *Nyl.*
virens, *Tuck.*

43. Sagedia (*Mass.*), *Kbr.*
1. chlorotica (*Ach.*), *Mass.*
2. oxyspora (*Nyl.*), *Tuck.*

44. Verrucaria, *Pers.*
maura (*Wahl.*), *Th. Fr.*

45. Pyrenula (*Ach.*), *Næg. & Hepp.*
1. punctiformis (*Ach.*), *Næg.*
2. gemmata (*Ach.*), *Næg.*
3. glabrata (*Ach.*), *Mass.*
4. nitida, *Ach.*

FUNGI.

The representatives of this very extensive and difficult family of plants are of course widely distributed in all parts of the county, but owing to the impossibility of preparing herbarium specimens of the softer species which would be of any value, and to the fact that so few persons are interested in the study of the subject, no collection has ever been brought together of sufficient size to make the enumeration of the species of any service as a guide to the study of the Fungi of the county. A very full list of the Fungi growing in the vicinity of Amherst, Mass., particularly those about Brattleboro, Vermont, was prepared by the late Charles C. Frost of the last-named place, and will be found in the "Catalogue of Plants growing within thirty miles of Amherst College," published in 1875. This list probably includes many species not found here, but can be used generally to assist the collector to the species which may be looked for in this region, and the reader is also referred to the "List of Fungi collected in the vicinity of Boston" (Proc. Boston Soc. Nat. Hist., Vol. VI, 1856), and the "List of Fungi found in the vicinity of Boston" (Prof. W. G. Farlow, in Bull. Bussey Inst., March, 1876, and Jan., 1878) for further assistance. Moreover, it is probable, as has been suggested by an eminent authority, that without a complete knowledge of European forms from living specimens, and a familiarity with the foreign literature upon the subject, very little original work can be accomplished, at least to be of permanent value.

MARINE ALGÆ.

(Sea Weeds.)

Geographically considered, Essex county waters extend from, and include, Salisbury beach on the north to Saugus river on the south, but in more general terms the region from Boston northward to Portsmouth may be considered as representing our marine flora as a more natural division; or, the north shore of Massachusetts Bay and the north shore of Cape Ann, or Ipswich Bay, and the sand beaches northward. The flora of this region is an interesting one. Cape Cod projecting its long arm into the sea seems to form a natural barrier to the progress of southern species northward, for here the warm current of the Gulf stream bears off to the eastward, while between it and the shore north of Cape Cod, the almost expended influences of the cold Labrador current are slightly felt. A very marked separation is therefore made between the Arctic and southern marine flora and fauna making our waters quite Arctic in their character. Yet in warm places along the shore a few daring southern species have taken up their abode, while some northern ones are to be found south of Cape Cod, besides the species termed cosmopolitan which flourish both north and south. Hence, we find the Essex county waters to be the northern limit of many species and the southern limit of others, therefore, it becomes very desirable to give as far as possible the precise localities of specimens collected. This list has been prepared from the private lists of Mrs. M. H. Bray of Gloucester, Rev. A. B. Hervey of Taunton, who has passed his summer vacations of late at Marblehead Neck, Mr. Frank S. Collins[3] of Malden, whose careful collecting has extended from Nahant to Cape Ann, the herbarium of the Peabody Academy of Science, and the papers by Prof. W. G. Farlow of Harvard College, in Proceedings of the American Academy of Arts and Sciences, Vol. X, 1875, and in the Report of the U. S. Fish Commission for 1876. Prof. Farlow, although his time is fully occupied with more important matters, has most kindly reviewed the writer's manuscript relating to the Marine Algæ, offering valuable suggestions and adding several notes which will be found below. It is thought that this list is quite complete, lacking only those species which have not as yet been published, or are not yet fully determined, of very recent collection.

---

[3] Mr. Collins has kindly revised the list to the present date, November, 1880.

ALGÆ.

FLORIDEÆ.

RHODOMELEÆ (including Laurencieæ).

Dasya elegans, *Ag.* (CHENILLE.)
Reported from Salem and Ipswich, probably with insufficient authority. The range of this species is from "Key West to Cape Cod" (Dr. Farlow's list).

Bostrychia rivularis, *Harv.*
"Isles of Shoals, N. H. (?) to Florida" (Dr. Farlow's list). This is doubtful as an Essex county species.

Polysiphonia urceolata, *Grev.*
Common.
Var. **patens.**
Cape Ann (Dr. Farlow's list).
Var. **formosa.**
New England (Dr. Farlow's list).

Polysiphonia subtilissima, *Mont.*
Essex county (Rev. A. B. Hervey), Newburyport (Dr. Farlow's list), Gloucester (Dr. Farlow). More common southward.

Polysiphonia Olneyi, *Harv.* (DOUGH-BALLS.)
Essex county (Rev. A. B. Hervey).

Polysiphonia Harveyi, *Bailey.* (NIGGER-HAIR.)
Salem (Rev. A. B. Hervey), Gloucester (Dr. Farlow, Mrs. Bray).
Var. **arietina,** *Harv.*
Little Nahant (Dr. Farlow), Lynn beach (Frank S. Collins).

Polysiphonia elongata, *Grev.* (LOBSTER-CLAWS).
Lynn, southward (Dr. Farlow's list). Frequent along the shore from Cape Ann, southward (Rev. A. B. Hervey and Frank S. Collins).

Polysiphonia violacea, *Grev.*
Salem, Gloucester (Mrs. Bray, Frank S. Collins); Ipswich beach, a most beautiful specimen (Mrs. T. S. Greenwood).

Polysiphonia fibrillosa, *Grev.*
Cape Ann, southward (Dr. Farlow, in herb. P. A. S.); Essex county waters (Rev. A. B. Hervey, Mrs. Bray).

Polysiphonia variegata, *Ag.*
Danvers (Rev. A. B. Hervey, in herb. P. A. S.).

Polysiphonia atrorubescens, *Grev.*
Gloucester (Mrs. A. L. Davis, in herb. P. A. S.); frequent from Gloucester, southward (Rev. A. B. Hervey, Mrs. Bray).

Polysiphonia nigrescens, *Grev.*
Frequent.
Polysiphonia fastigiata, *Grev.*
Common.
Rhodomela subfusca, *Ag.*
Little Nahant (Dr. Farlow, in herb. P. A. S.). "Common from Boston northward" (memo. Dr. Farlow).
Var. gracilis.
Swampscott (Frank S. Collins), Gloucester (Mrs. A. L. Davis, Mrs. Bray).
Var. Rochei.
Salem (in herb. P. A. S.), Gloucester (Mrs. Bray).

SPHÆROCOCCOIDEÆ.

Delesseria sinuosa, *Lmx.*
Frequent.
Delesseria alata, *Lmx.*
Nahant, northward. Frequent.
Delesseria angustissima, *Griff.*
Gloucester (Dr. Farlow's list, Mrs. Bray, etc.).
Calliblepharis ciliata, *Kütz.*
Gloucester (Frank S. Collins), northward.
Gracilaria multipartita, *Ag.*
East coast (Dr. Farlow's list).

CORALLINEÆ.

Corallina officinalis, *L.*
Common.
Melobesia membranacea, *Lmx.*
Frequent (Rev. A. B. Hervey, Frank S. Collins).
Melobesia Lejolisii, *Rosanoff.*
Common at Nahant (Frank S. Collins). Recently separated from the last. This is the common species on eel-grass.
Melobesia farinosa, *Lmx.*
East coast (Dr. Farlow's list); Marblehead (Rev. A. B. Hervey).
Melobesia pustulata, *Lmx.*
Frequent (Frank S. Collins).
Lithothamnion polymorphum, *Aresch.*
Frequent (Frank S. Collins). Northern.
Hildenbrandtia rosea, *Kütz.*
Dredged along the coast, growing on stones; "everywhere between tide marks" (memo. Dr. Farlow).

RHODYMENIEÆ.

Rhodymenia palmata, *Grev.* (COMMON DULSE.)
Common.

Euthora cristata, *Ag.*
Nahant, northward.

Lomentaria rosea, *Thuret.* (Chylocladia rosea, *Harv.*)
Washed on the beach at Newport, R. I., Gay Head, Mass.; also found at Portsmouth, N. H. This ought to be collected in Essex county waters.

Rhabdonia tenera, *Ag., Bidrag.* (Solieria chordalis, *Ner. Am. Bor.*).
North side of Cape Ann; "only at Goose Cove, Squam, and the creek leading into it" (Dr. Farlow).

SPONGIOCARPEÆ.

Polyides rotundus, *Ag.*
Frequent.

SQUAMARIEÆ.

Petrocelis cruenta, *Ag.*
Nahant, Mass. (Dr. Farlow's list), also noticed by Frank S. Collins and Rev. A. B. Hervey. This is a northern species.

HELMINTHOCLADIEÆ.

Nemalion multifidum, *Ag.*
Frequent.

GIGARTINEÆ.

Phyllophora Brodiæi, *Ag.*
Quite common.

Phyllophora membranifolia, *Ag.*
Frequent.

Gymnogongrus Norvegicus, *Ag.* (inc. G. Torreyi, *Ag.*)
Beverly and Nahant (Dr. Farlow's list), Gloucester (Mrs. Bray).

Ahnfeltia plicata, *Fr.*
Frequent.

Cystoclonium purpurascens, *Kütz.*
Common.

Gigartina mamillosa, *Ag.*
Frequent.

Chondrus crispus, *Lyngb.* (IRISH MOSS.)
Very common.

DUMONTIEÆ.

Halosaccion ramentaceum, *Ag.*
Gloucester (Mrs. H. A. Cochrane, in herb. P. A. S.), northward.

SPYRIDIEÆ.

Spyridia filamentosa, *Harv.*
Massachusetts Bay, southward (Dr. Farlow's list).

CERAMIEÆ.

Ceramium rubrum, *Ag.*
Very common.

Ceramium Deslongchampsii, *Ch.*
Frequent. The specimens noted as C. Hooperi, *Harv.*, belong to this species.

Ceramium diaphanum, *Roth.*
Occasionally found on the New England coast (Dr. Farlow's list), Gloucester (Mrs. A. L. Davis, in herb. P. A. S.), Nahant (Frank S. Collins).

Ceramium strictum, *Harv.*
Gloucester (Mrs. Bray), etc. Not rare.

Ceramium fastigiatum, *Harv.*
Massachusetts Bay (Dr. Farlow's list).

Ptilota plumosa, *Ag.*, var. **serrata.**
Frequent from Nahant, northward.

Ptilota elegans, *Bonnem.*
Common.

Gloiosiphonia capillaris, *Carm.*
Cape Ann, southward.

Griffithsia corallina(?), *Ag.*
Gloucester (Rev. A. B. Hervey), southward. "G. corallina does not occur in N. A. South of Cape Cod is G. Bornetiana, Farlow. Have you specimens from Essex county? I have never seen anything but fragments of a Griffithsia" (memo. Dr. W. G. Farlow). Collectors will confer a favor by communicating specimens of this, or of any species outside of the usual localities, to the writer, for Dr. Farlow.

Callithamnion tetragonum, *Ag.*
Niles beach, Gloucester (Mrs. Bray). More frequent south.

Callithamnion Baileyi, *Harv.*
Lynn beach (Frank S. Collins). More frequent south.

Callithamnion byssoideum, *Arn.*
"Nahant" (memo. Rev. A. B. Hervey), "Nahant to New York" (Dr. Farlow's list).

Callithamnion corymbosum, *Ag.*
Frequent.

Callithamnion Americanum, *Harv.*
Frequent.

Callithamnion Pylaisæi, *Mont.*
Quite common.

Callithamnion floccosum, *Ag.*
Massachusetts Bay, northward (Dr. Farlow's list); Gloucester (Mrs. Bray).

Callithamnion Rothii, *Lyngb.*
This is not a true Callithamnion (memo. Dr. Farlow); growing on rocks, Gloucester, etc. (Frank S. Collins, Rev. A. B. Hervey).

PORPHYREÆ.

Porphyra vulgaris, *Ag.* (LAVER.)
Very common.

Bangia fuscopurpurea, *Lyngb.*
Essex county waters (Mrs. Bray, Rev. A. B. Hervey). Common.

(?) FLORIDEÆ.

(incertæ sedis.)

Chantransia Daviesii, *Thuret.*
Gloucester (Mrs. J. T. Lusk), also noticed by Rev. A. B. Hervey and Mrs. Bray; Gay Head (Dr. Farlow).

Chantransia virgatula, *Thuret.*
New York, northward (Dr. Farlow's list); Gloucester (Mrs. Bray).

Erythrotrichia ceramicola, *Aresch.*
Cape Ann (Mrs. Bray, Rev. A. B. Hervey); Buzzard's Bay; Cape Ann; Portland harbor, Me. (Dr. Farlow's list).

Goniotrichum elegans, *Zanard.*
On Dasya. Gloucester, Mass., collected by Mrs. J. T. Lusk. (Dr. Farlow, in Proc. Am. Acad., Vol. X, p. 351).

MELANOSPORÆ.

FUCACEÆ.

Fucus nodosus, *L.* (ROCK-WEED.)
Frequent.

Fucus distichus, *L.* (F. filiformis, *Gm.*)
Marblehead (Dr. Farlow's list, Frank S. Collins), also collected by Mrs. Bray.

Fucus furcatus, *Ag.*
Nahant, northward (Dr. Farlow's list), Marblehead, etc. (Frank S. Collins, Mrs. Bray).

Fucus ceranoides, *L.*
East coast (Dr. Farlow's list), also Rev. A. B. Hervey. This is somewhat doubtful as a county species.

Fucus vesiculosus, *L.* (ROCK-WEED.)
Common.

Fucus serratus, *L.*
Newburyport, Mass., northward (Dr. Farlow's list). This may prove to have been introduced by shipping (Frank S. Collins, letter Nov. 20, 1880).

Fucus evanescens, *Ag.*
Abundant on the coast of Maine, will probably be found in Essex county (Frank S. Collins, letter Nov. 20, 1880).

PHÆOSPOREÆ.

Alaria esculenta, *Grev.*
Frequent.

Laminaria dermatodea, *De la Pyl.*
Cape Ann (Mrs. Bray), from Marblehead, northward (memo. Dr. Farlow).

Laminaria saccharina, *Lmx.*
Common.

Laminaria longicruris, *De la Pyl.*
Also common. Both this and the last go by the name, "Devil's Apron," or "Kelp," as also does the next usually.

Laminaria flexicaulis, *Le Jolis.* (L. digitata, *Lam.*, in part.)
Frequent (Frank S. Collins, Mrs. Bray).

Laminaria platymeris, *De la Pyl.*
Cape Ann (Mrs. Bray), New England(?) (Dr. Farlow's list), Swampscott (Frank S. Collins).

Agarum Turneri, *Post & Rupr.* (SEA-COLANDER)
Nahant, northward. Common.

Asperococcus compressus, *Griff.*
Gloucester (Dr. Farlow's list), also collected by Mrs. Bray and Rev. A. B. Hervey.

Asperococcus echinatus, *Grev.*
Cape Ann (Rev. A. B. Hervey, Mrs. Bray). Frequent along the New England coast.

Ralfsia verrucosa, *Aresch.*
Frequent from Nahant, northward.
Ralfsia clavata, *Crouan.*
Swampscott; Malden marshes, southern limit (Frank S. Collins). Rare. Common, northward (memo. Dr. Farlow).
Chorda filum, *Stack.*
Common.
Chordaria flagelliformis, *Ag.*
Frequent.
Chordaria divaricata, *Ag.*
Swampscott (Rev. A. B. Hervey), Gloucester (Mrs. Bray), north side of Cape Ann (Dr. Farlow), southward to New York.
Castagnea virescens, *Thuret.* (Mesogloia virescens, *Harv.*, *Ner. Am. Bor.*)
Frequent at Gloucester, etc.
Castagnea Zosteræ, *Thuret.* (Mesogloia Zosteræ, *Harv.*, *Ner. Am. Bor.*)
Gloucester (Frank S. Collins, Mrs. Bray), Wood's Hole (Dr. Farlow's list).
Leathesia tuberiformis, *Gray.*
Frequent.
Elachista fucicola, *Fr.*
Frequent.
Myrionema strangulans, *Grev.*
Essex county waters (Rev. A. B. Hervey, Frank S. Collins).
Cladostephus spongiosus, *Ag.*
Gloucester, etc. (Mrs. Bray).
Cladostephus verticillatus, *Ag.*
Occasional (Rev. A. B. Hervey, Mrs. Bray).
Sphacelaria radicans, *Ag.*
Nahant (Dr. Farlow), Gloucester, etc. (Mrs. Bray, Rev. A. B. Hervey).
Sphacelaria cirrhosa, *Ag.*
New York, northward (Dr. Farlow's list), Gloucester (Mrs. Bray).
Ectocarpus sphærophorus, *Carm.*
"Nahant. I found a very small quantity of this last summer (1879), and it is, I believe, all that has been found in this country" (Frank S. Collins, in letter Nov. 24, 1879).
Ectocarpus brachiatus, *Harv.*
Boston, northward (Dr. Farlow's list), also reported in county waters (Rev. A. B. Hervey), but Dr. Farlow now considers this as doubtful.

Ectocarpus firmus, *Ag.* (E. littoralis, *Harv.*)
Rockport (in herb. P. A. S.), and elsewhere (Rev. A. B. Hervey, Frank S. Collins).

Ectocarpus Farlowii, *Thuret.*
Marblehead, northward (Dr. Farlow's list, Mrs. Bray, and Rev. A. B. Hervey).

Ectocarpus siliculosus, *Lyngb.*
Frequent.

Ectocarpus viridis, *Harv.*
Quite frequent.

Ectocarpus tomentosus, *Lyngb.*
From Boston, northward; not very frequent.

Ectocarpus fasciculatus, *Harv.*
Frequent.

Ectocarpus granulosus, *Ag.*
Boston harbor (Dr. Farlow's list), also noticed by Rev. A. B. Hervey.

Ectocarpus Durkeei, *Harv.*
Portsmouth, N. H. (Dr. Farlow's list). This species ought to be found in Essex county. It is probably only a form of the preceding species (memo. Dr. Farlow).

Dictyosiphon foeniculaceus, *Grev.*
Frequent.

Desmarestia aculeata, *Lmx.*
Common.

Desmarestia viridis, *Lmx.*
Common.

Punctaria latifolia, *Grev.*
Frequent.

Var. **Zosteræ,** *Le Jolis.* (P. tenuissima, *Harv.*)
Gloucester (Mrs. Bray). Common at Lynn beach in early spring (Frank S. Collins).

Punctaria plantaginea, *Grev.*
Frequent.

Phyllitis Fascia, *Ktz.* (Laminaria Fascia, *Ag.*)
Common.

Scytosiphon lomentarius, *Ag.* (Chorda lomentaria, *Lyngb.*)
Quite common.

CHLOROSPORÆ.

SIPHONEÆ.

Bryopsis plumosa, *Lmx.*
Common.

Vaucheria piloboloides, *Thuret.*
"Undoubtedly in Essex county. It is very abundant in the salt marshes at Malden. There are other Vaucheriæ along the shore, but I do not think the species are definitely settled" (Frank S. Collins, in letter Nov. 24, 1879).

ZOÖSPOREÆ.

Enteromorpha intestinalis, *Link.*
Very common.
Enteromorpha compressa, *Grev.*
Very common.
Enteromorpha clathrata, *Grev.*
Frequent.
Ulva latissima, *Linn.* (SEA-LETTUCE; SEA-CABBAGE.)
Very common. This, and sometimes the next, are the plants which so abound in brackish waters along our coast. When decaying in large quantities, the Beggiatoæ[4] upon them produce the well known disagreeable smell so familiar to those residing near tide ponds and flats. This may be given as V. lactuca in future works.
Ulva lactuca, *Linn.* (Monostroma Grevillei, *Le Jolis*).
Common. Marblehead (Frank S. Collins), Gloucester (Mrs. Bray), etc.
Cladophora rupestris, *Linn.*
Common.
Cladophora arcta, *Dillw.*
Quite frequent.
Cladophora lanosa, *Roth.*
Nahant (Frank S. Collins, Dr. Farlow), also noticed by Mrs. Bray.
Cladophora uncialis, *Fl. Dan.*
Occasional (Rev. A. B. Hervey, Frank S. Collins).
Cladophora glaucescens, *Griff.*
Nahant (Frank S. Collins), also noticed at Magnolia by Mrs. Bray. A more southern species.
Cladophora flexuosa, *Griff.*
Nahant (Frank S. Collins), etc. Occasional.
Cladophora refracta, *Roth.*
Marblehead (Frank S. Collins, Rev. A. B. Hervey), Gloucester (Mrs. Bray). Southern.
Cladophora albida, *Huds.*
Gloucester (Mrs. Bray). Southern.

[4] A low parasitic alga (see Farlow, Bull. Bussey Inst., Jan., 1877, p. 76).

Cladophora gracilis, *Griff.*
Nahant (Frank S. Collins, Dr. Farlow's list), to N. J.

Cladophora lætevirens, *Dillw.*
Boston Bay (Dr. Farlow's list), Nahant (Frank S. Collins), also at New York.

Cladophora fracta, *Fl. Dan.*
Nahant (Frank S. Collins), also noticed by Rev. A. B. Hervey. Common.

Cladophora expansa, *Kütz.*
"Not uncommon at Nahant and Marblehead" (Frank S. Collins, in letter Nov. 24, 1879).

Chætomorpha Piquotiana, *Mont.*
Frequent (Rev. A. B. Hervey, Frank S. Collins).

Chætomorpha melagonium, *Web. & Mohr.*
Frequent.

Chætomorpha ærea, *Dillw.*
East coast (Dr. Farlow's list), Gloucester (memo. Dr. Farlow).

Chætomorpha sutoria, *Berk.*
Marblehead (Frank S. Collins), also noticed at Magnolia, Mass., by Mrs. Bray. More abundant southward.

Chætomorpha tortuosa, *Dillw.*
Nahant, northward. Common.

Hormotrichum Younganum, *Dillw.*
Gloucester (Mrs. H. A. Cochrane, in herb. P. A. S.), Nahant (Frank S. Collins). Frequent.

Hormotrichum speciosum, *Carm.*
Swampscott (Frank S. Collins).

Hormotrichum collabens, *Kütz.*
"Nahant. This last is an extremely rare species, having been found only once in Great Britain, once in France, and a very few times in the German ocean. I have seen it only once. Both the Hormotrichum (the last and this species) were identified by Dr. E. Bornet of Paris, from my specimens" (Frank S. Collins, in letter Nov. 24, 1879).

Hormotrichum Carmichaelii, *Harv.*
Swampscott (Frank S. Collins).

CYANOPHYCEÆ.

Lyngbya ferruginea, *Ag.*
(Rev. A. B. Hervey, Mrs. Bray). Very common (memo. Dr. Farlow).

Calothrix confervicola, *Ag.*
Common.

Calothrix scopulorum, *Ag.*
Common.

Calothrix crustacea, *Thuret.*
Nahant (Frank S. Collins).

Rivularia atra, *Roth.*
New England (Dr. Farlow's list).

Clathrocystis roseo-persicina, *Cohn.*
New England (Dr. Farlow's list), Gloucester (Mrs. Bray).

ADDITIONAL SPECIES.

Mesogloia vermicularis, *Ag.*
"Occasionally found at Cape Ann" (Dr. Farlow, in Proc. Am. Acad., Vol. X), Gloucester (Mrs. A. L. Davis, in herb. P. A. S.).

Rhizoclonium Kochianum, *Kütz.*
"Lynn beach, and since found at Gloucester" (Frank S. Collins, in letter of Nov. 24, 1879).

Rhizoclonium riparium, *Harv.*
This will be found in Essex county, as it belongs to this region.

Monostroma pulchrum, *Farlow,* n. sp.
Nahant (Frank S. Collins).

NOTE.—The fresh-water Algæ of the county have not received sufficient attention from any one to make it possible to present here a list even of the larger and more conspicuous species. Like the Fungi, they contain immense numbers of microscopic species, many of which have not been studied enough to be arranged in their proper places. Among the lower forms are the Diatoms, which are also found in salt water, and perhaps it is well to mention that the Protococcus nivalis (or Red Snow of the Arctic Voyagers) has been observed at Nahant.

ADDENDA.

Hon. Nathaniel Silsbee, formerly of Salem, has kindly furnished the following note relating to Rev. Manasseh Cutler:

Mr. Silsbee's father, Nathaniel Silsbee, attended Dr. Cutler's school, and, with Willard Peele, harnessed the vehicle in which Dr. Cutler went to Ohio. Dr. Cutler brought with him, on his return, a tulip tree which he planted in his garden at Hamilton. It is said that the tulip tree grows wild in the Magnolia swamp.

This last statement has been made by others, although no specimens of the tree have ever been seen from that locality. It is possible that seedlings may have originated from Dr. Cutler's tree, but even this is uncertain.

Page 34. **Alyssum montanum,** *L.*
Danvers, 1880 (J. H. Sears). (Nat. from Eu.) Not previously noticed.

" 40. **Claytonia Caroliniana,** *Mx.*,
is the species evidently intended by those who collected the specimens, as that is the one found in this region. Both are, however, included in one species by some authors.

" " **Malva crispa,** *L.* (CURLED MALLOW.)
Salem, 1880 (J. R.). (Adv. from Eu.) Not previously noticed.

" 41. **Linum striatum,** *Walt.*
Gloucester, 1880 (J. H. Sears). Not previously noticed.

" 60. **Dipsacus sylvestris,** *Mill.* (TEASEL.)
Add, Rockport (Mrs. Wheeler). (Adv. from Eu.)

" 64. **Iva xanthiifolia,** *Nutt.*
Quite abundant around the old carpet factory at Tapleyville (J. H. Sears). (Adv. from the west.) Not previously noticed.

" 71. **Gaylussacia dumosa,** *T. & G.*
Add, Cedar and Long ponds, Saugus (Herbert A. Young).

" 109. **Cypripedium pubescens,** *Willd.*
Add, West Boxford! (Mrs. Wilmarth).

Page 121. **Carex Pseudo-Cyperus,** *L.*
Middleton, Aug., 1880 (J. R.). Not previously noticed.

The following list of mosses was overlooked at the time of preparing the flora. It is interesting, as being the first list of mosses, of this vicinity, to appear in print. It is taken from a more extended list of "Musci of Eastern Massachusetts, by John Lewis Russell. Read before the Nat. Hist. Soc., Dec. 4, 1844" (Boston Journal of Natural History, Vol. V, 1845-7). It is printed as given in the original, the modern names being added in parentheses when required. An asterisk is placed against those species which have not been previously noticed.

* **Sphagnum squarrosum,** *W. & M.*
Tewksbury (B. D. Greene).

Sphagnum obtusifolium, *Ehr.* (S. palustre.)
Manchester (Oakes).

Sphagnum acutifolium, *Ehr.*
Magnolia swamp, Manchester (Oakes).

Gymnostomum truncatulum, *Hoffm.* (Pottia truncata.)
Sandy garden walks. October, Salem (Russell).

Gymnostomum pyriforme, *Hedw.* (Physcomitrium.)
Vicinity of Salem (Russell).

Anictangium ciliatum, *Hedw.* (Hedwigia ciliata.)
Common on bowlders, rocks, and stone walls.

Diphyscium foliosum, *Mohr.*
In profuse abundance on a hard-trodden path in woods at Danvers, literally affording a carpet of bristly perichætia. It occurs in abundance about Ipswich (Oakes).

Tetraphis pellucida, *Hedw.*
Ipswich (Russell).

* **Splachnum ampullaceum,** *L.*
Manchester (Oakes).

Grimmia maritima, *Trin.* (Schistidum.)
Nahant (Greene).

Grimmia apicola, *Sw.* (Schistidium apocarpa.)
Nahant (Greene).

* **Didymodon capillaceus,** *Schrad.* (Distichium.)
In perfect fruit in January, Salem (Russell). (Locality doubtful.)

Dicranum glaucum, *Hedw.* (Leucobryum vulgare.)
Manchester (Oakes).

Dicranum flagellare, *Hedw.*
Ipswich (Oakes).
Dicranum scoparium, *Hedw.*
Salem, Ipswich, etc. (Russell).
* **Dicranum rugosum,** *Dill.* (D. undulatum.)
Ipswich (Oakes).
Dicranum strictum, *Schl.*
Ipswich (Oakes). (Undoubtedly D. flagellare.)

Polytrichum angustatum, *Brid.* (Atrichum.)
Russell, Jour. Essex Co. Nat. Hist. Soc., p. 92.
* **Polytrichum gracile,** *Menzies.*
Boggy places, Ipswich (Oakes).
Polytrichum commune, *L.*
Common.
Polytrichum piliferum, *Schreb.*
Common.
* **Polytrichum Pennsylvanicum,** *Hedw.* (Pogonatum brevicaule, *Brid.*)
Russell, Jour. Essex Co. Nat. Hist. Soc., under P. boreale.

Funaria hygrometrica, *Hedw.*
Danvers; garden paths, Salem (Russell).

Orthotrichum saxatile, *Brid.* (O. anomalum.)
Vicinity of Salem (Russell).
* **Orthotrichum affine,** *Schrad.*
Common on the bark of trees. (Locality doubtful.)
Orthotrichum strictum, *Brid.* (O. Hutchinsiæ.)
Ipswich (Oakes), on rocks.
* **Orthotrichum crispum,** *Hedw.*
Ipswich (Oakes).
* **Orthotrichum clavellatum,** *Dell.* (Drummondia clavellata, *Hk.*)
Ipswich (Oakes).

* **Bryum argenteum,** *L.*
Common.
Bryum cæspititium, *L.*
Common.
* **Bryum roseum,** *Schreb.*
Tewksbury (Greene).
Bryum punctatum, *Schreb.* (Mnium.)
Ipswich (Oakes).
Bryum cuspidatum, *L.* (Mnium.)
Ipswich (Oakes).
Bryum hornum, *L.* (Mnium.)
Tewksbury (Greene).

Bartramia longifolia, *Hk.*
Crevices of shady rocks. Common. (Undoubtedly B. pomiformis.)

Bartramia fontana, *Sw.*
On rocks, over which water trickles, Danvers (Russell).

Bartramia crispa, *Sw.* (B. pomiformis.)
Ipswich (Oakes).

Pterigynandrum intricatum, *Hedw.* (Pylaisæa.)
On apple trees, etc. Ipswich (Oakes).

* **Leucodon brachypus,** *Brid.*
Ipswich (Oakes), Tewksbury (Greene).

Neckera pennata, *Hedw.*
Ipswich (Oakes).

Climacium Americanum, *Rich.*
Ipswich (Oakes).

Leskea compressa, *Hedw.*
Ipswich (Russell). (This is probably Cylindrothecium cladorrhizans.)

Leskea polycarpa, *Ehr.*
Ipswich (Oakes).

* **Hypnum salebrosum,** *Hoff.*
Salem, common (Russell).

Hypnum splendens, *Hedw.*
Ipswich (Oakes).

Hypnum Schreberi, *Willd.*
Manchester (Oakes).

Hypnum catenulatum, *Brid.*
Near Marblehead (Russell). (Probably H. adnatum.)

Hypnum triquetrum, *L.*
Ipswich (Russell).

Hypnum cupressiforme, *L.*
Ipswich (Russell).

* **Hypnum serrulatum,** *Muhl.*
Manchester (Oakes).

Hypnum imponens, *Muhl.*
Ipswich (Russell).

Hypnum crista-castrensis, *L.*
Manchester (Oakes).

Page 114. **Hypnum campestre.**
This is only a form of H. velutinum.

SUMMARY.

	ORDERS.	GENERA.	SPECIES.	VARIETIES.	FOREIGN INTRODUCED.	U. S. INTRODUCED.
1	Ranunculaceæ.	9	29	1	3	.
2	Magnoliaceæ.	2	2	.	.	1
3	Berberidaceæ.	3	3	.	1	.
4	Nymphæaceæ.	2	3	1	.	.
5	Sarraceniaceæ.	1	1	.	.	.
6	Papaveraceæ.	3	3	.	2	.
7	Fumariaceæ.	4	4	.	1	1
8	Cruciferæ.	14	26	3	16	.
9	Violaceæ.	1	11	1	2	.
10	Cistaceæ.	3	7	.	.	.
11	Droseraceæ.	1	2	.	.	.
12	Hypericaceæ.	2	7	.	1	.
13	Elatinaceæ.	1	1	.	.	.
14	Caryophyllaceæ.	10	27	.	13	.
15	Paronychieæ.	2	2	.	1	.
16	Ficoideæ.	1	1	.	.	1
17	Portulacaceæ.	2	2	.	1	.
18	Malvaceæ.	4	8	.	6	.
19	Tiliaceæ.	1	2	.	1	.
20	Linaceæ.	1	2	.	1	.
21	Geraniaceæ.	4	8	.	.	.
22	Rutaceæ.	2	2	.	1	1
23	Anacardiaceæ.	1	5	.	.	.
24	Vitaceæ.	2	5	.	.	.
25	Rhamnaceæ.	2	2	.	1	.
26	Celastraceæ.	1	1	.	.	.
27	Sapindaceæ.	3	7	.	3	1
28	Polygalaceæ.	1	5	.	.	.
29	Leguminosæ.	17	39	.	12	5
30	Rosaceæ.	12	52	3	.	.
31	Saxifragaceæ.	6	11	1	3	2
32	Crassulaceæ.	3	6	.	4	1
33	Hamamelaceæ.	1	1	.	.	.
34	Halorageæ.	2	4	3	.	.
35	Onagraceæ.	4	12	3	1	.
36	Melastomaceæ.	1	1	.	.	.
37	Lythraceæ.	3	4	.	.	.
38	Cactaceæ.	1	1	.	.	1
39	Cucurbitaceæ.	2	2	.	.	2
40	Umbelliferæ.	16	20	.	6	.
41	Araliaceæ.	2	5	.	1	.

Summary. (Continued.)

	ORDERS.	GENERA.	SPECIES.	VARIETIES.	FOREIGN INTRODUCED.	U. S. INTRODUCED.
42	Cornaceæ.	2	8	.	.	.
43	Caprifoliaceæ.	7	14	1	.	2
44	Rubiaceæ.	4	8	1	.	.
45	Dipsaceæ.	1	1	.	1	.
46	Compositæ.	43	129	7	29	5
47	Lobeliaceæ.	1	5	.	.	.
48	Campanulaceæ.	2	5	.	2	.
49	Ericaceæ.	18	35	2	.	.
50	Ilicineæ.	2	5	.	.	.
51	Plantaginaceæ.	1	4	.	1	.
52	Plumbaginaceæ.	1	1	.	.	.
53	Primulaceæ.	6	11	1	3	.
54	Lentibulaceæ.	1	7	.	.	.
55	Bignoniaceæ.	3	3	.	.	3
56	Orobanchaceæ.	2	2	.	.	.
57	Scrophulariaceæ.	14	29	.	7	1
58	Verbenaceæ.	2	3	.	.	.
59	Labiatæ.	22	35	.	17	1
60	Borraginaceæ.	7	10	.	7	.
61	Polemoniaceæ.	1	1	.	.	.
62	Convolvulaceæ.	3	6	.	3	.
63	Solanaceæ.	8	12	.	10	2
64	Gentianaceæ.	4	5	.	.	.
65	Apocynaceæ.	1	2	1	.	.
66	Asclepiadaceæ.	1	7	.	.	.
67	Oleaceæ.	3	5	.	2	.
68	Aristolochiaceæ.	1	1	.	.	1
69	Phytolaccaceæ.	1	1	.	.	.
70	Chenopodiaceæ.	6	15	2	6	.
71	Amarantaceæ.	1	3	.	3	.
72	Polygonaceæ.	3	25	2	8	.
73	Lauraceæ.	2	2	.	.	.
74	Thymeleaceæ.	1	1	.	1	.
75	Santalaceæ.	1	1	.	.	.
76	Ceratophyllaceæ.	1	1	.	.	.
77	Callitrichaceæ.	1	1	.	.	.
78	Euphorbiaceæ.	2	7	.	2	1
79	Urticaceæ.	9	12	.	5	3
80	Platanaceæ.	1	1	.	.	1
81	Juglandaceæ.	2	7	.	1	1
82	Cupuliferæ.	6	15	1	1	.
83	Myricaceæ.	2	3	.	.	.
84	Betulaceæ.	2	8	.	.	.
85	Salicaceæ.	2	21	2	7	1
86	Coniferæ.	7	17	.	4	3

Summary. (Concluded.)

	ORDERS.	GENERA.	SPECIES.	VARIETIES.	FOREIGN INTRODUCED.	U. S. INTRODUCED.
87	Araceæ.	5	5	.	.	.
88	Lemnaceæ.	1	2	.	.	.
89	Typhaceæ.	2	4	2	.	.
90	Naiadaceæ.	5	24	4	1	.
91	Alismaceæ.	5	7	7	.	.
92	Hydrocharidaceæ.	2	2	.	.	.
93	Orchidaceæ.	13	31	1	.	.
94	Amaryllidaceæ.	1	1	.	.	.
95	Iridaceæ.	2	3	.	.	.
96	Smilaceæ.	1	3	.	.	.
97	Liliaceæ.	18	27	.	4	1
98	Juncaceæ.	2	14	2	.	.
99	Pontederiaceæ.	2	2	.	.	.
100	Xyridaceæ.	1	1	1	.	.
101	Eriocaulonaceæ.	1	1	.	.	.
102	Cyperaceæ.	9	111	9	2	1
103	Gramineæ.	50	121	7	33	4
104	Lycopodiaceæ.	3	9	2	1	.
105	Rhizocarpeæ.	1	1	.	.	.
106	Ophioglossaceæ.	2	5	4	.	.
107	Equisetaceæ.	1	4	.	.	.
108	Filices.	13	29	11	.	.
109	Musci.	33	115	8	.	.
110	Hepaticæ.	19	25	4	1	.
111	Nitelleæ.	1	4	1	.	.
112	Chareæ.	1	4	2	.	.
113	Lichens.	45	157	35	.	.
114	Fungi(species not studied.)					
115	Algæ (Marine).	70	151	6	.	.

	Orders.	Genera.	Species.	Varieties.	U. S. Introduced.	Foreign Introduced.	Woody Plants Native.	Woody Plants U. S. Int.	Woody Plants For. Int.
Exogens.	85	357	814	34	38	198	157	27	40
Gymnosperms.	1	7	17	.	3	4	10	3	4
Endogens.	17	120	359	33	6	40	1	.	.
	103	484	1190	67	47	242	168	30	44
Vasc. Cryptogams.	5	20	48	17	.	1	.	.	.
Muscineæ.	2	52	140	12	.	1	.	.	.
Characeæ.	2	2	8	3	.	.	,	.	.
Thallophytes.	3	115	308	41	.	.	.	.	.
	12	189	504	73	.	2	2	.	.
Total.	115	673	1694	140	47	246	168	30	44

ACKNOWLEDGMENT.

In the foregoing pages, the writer has endeavored to give proper credit to those who have contributed towards extending the size of the flora, or a knowledge of the distribution of the species. But other assistance has been rendered him by persons not directly interested in the local character of the work which should not pass without an acknowledgment. He is indebted to Prof. Asa Gray, Prof. G. L. Goodale, and Prof. Sereno Watson of the Cambridge herbarium, for much valuable assistance during the past five years in identifying many difficult species of plants; to Prof. W. G. Farlow, for many kindnesses in determining the lower cryptogams; to Prof. T. P. James, for similar services regarding the mosses; to Prof. C. S. Sargent, for much information of value concerning the trees and shrubs, and to Mr. Chas. J. Sprague, not only for his contribution of the entire list of lichens, but also for kindly remembering to communicate any rare plant found while collecting in Essex county.

The writer also desires to express his obligations to Messrs. Edwin and Charles E. Faxon, for their kindness in revising the final proof, and to Mr. Woodbury P. Conant, for his assistance in preparing the index.

INDEX.

Aaron's Rod, 52.
Abies Canadensis (Tsuga), 101.
Abies nigra, 12, 101.
Abies nigra (Picea), 101.
Abutilon, 40.
Acalypha, 94.
Acer, 43.
Achillea, 66.
Acknowledgment, 177.
Acolium, 155.
Acorus, 103.
Actæa, 31.
Adam-and-Eve, 109.
Adiantum, 133.
Adder's-mouth, 109.
Adder's-tongue Family, 132.
Adlumia, 32.
Æsculus, 43.
Æthusa, 56.
Agarum, 163.
Agrimonia, 48.
Agrimony, 48.
Agrostis, 123.
Ahnfeltia, 160.
Ailanthus, 42.
Aira, 128.
Alaria, 163.
Alder, 99.
Alectoria, 150.
Algæ, 158.
Alisma, 106.
Alismaceæ, 106.
Allium, 112.
Alnus, 99.
Alopecurus, 123.
Alsyke, 45.
Althæa, 40.

Alyssum, 34, 169.
Amarantaceæ, 90.
Amaranth Family, 90.
Amarantus, 90.
Amaryllidaceæ, 110.
Amaryllis Family, 110.
Ambrosia, 64.
Amelanchier, 51.
American Aspen, 100.
American Beech, 97.
American Cranberry, 71.
American Elm, 94.
American Holly, 74.
American Mountain Ash, 50.
American Pennyroyal, 81.
American White Hellebore, 111.
American Yew, 102.
Ammannia, 54.
Ampelopsis, 43.
Amphicarpæa, 47.
Anacharis, 107.
Anacardiaceæ, 42.
Anagallis, 76.
Andromeda, 72.
Andropogon, 130.
Anemone, 29.
Angelica, 56.
Anictangium, 170.
Anomodon, 141.
Antennaria, 67.
Anthoxanthum, 129.
Antirrhinum, 78.
Anychia, 39.
Aphyllon, 78.
Apios, 46.
Aplectrum, 109.
Apocynaceæ, 87.

Apocynum, 87.
Apple of Peru, 86.
Aquifoliaceæ, 74.
Aquilegia, 30.
Arabis, 33.
Araceæ, 103.
Aralia, 57.
Araliaceæ, 57.
Aramantus, 15, 91.
Archangelica, 56.
Arctostaphylos, 72.
Arenaria, 38.
Arethusa, 108.
Argemone, 32.
Arisæma, 103.
Aristida, 125.
Aristolochiaceæ, 89.
Arrhenatherum, 128.
Arrow Arum, 103.
Arrow-leaved Tear-thumb, 92.
Arrow-wood, 59.
Artemesia, 67.
Artemisia Absinthium, 15, 67.
Arthonia, 155.
Arum Family, 103.
Asarum, 89.
Asclepiadaceæ, 87.
Asclepias, 87.
Asparagus, 112.
Asperococcus, 163.
Asperugo, 83.
Aspidium, 134.
Asplenium, 134.
Aster, 61.
Atrichum, 140.
Atrichum (Polytrichum), 171.
Atriplex, 90.
Aulacomonion, 140.
Austrian Pine, 101.
Avena, 128.
Azalea (Rhododendron), 73.

Bæomyces, 154.
Bald Cypress, 102.

Ballota, 83.
Balm-of-Gilead, 100.
Balsam, 41.
Bangia, 162.
Baptisia, 47.
Barbarea, 33.
Barberry, 31.
Barberry Family, 31.
Barbula, 139.
Barley, 128.
Barn-yard Grass, 130.
Bartonia, 87.
Bartramia, 141, 171, 172.
Bartramieæ, 141.
Basket Osier, 99.
Bass-wood, 40.
Bayberry, 98.
Bayonet Rush, 113.
Beach Pea, 46.
Beach Plum, 47.
Beaked Hazelnut, 98.
Bearberry, 72.
Beard Grass, 123.
Bear Oak, 97.
Bedstraw, 59.
Beech, 97.
Beech-drops, 77.
Beech Fern, 134.
Beggar's-lice, 84.
Beggar-ticks, 66.
Bellflower, 70.
Bellwort, 111.
Belpharozia, 145.
Benjamin Bush, 93.
Berberidaceæ, 81.
Berberis, 31.
Betula, 98.
Betulaceæ, 98.
Biatora, 154.
Bidens, 66.
Bignoniaceæ, 77.
Bignonia Family, 77.
Bind-weed, 84.
Birch Family, 98.

Bird's-foot Violet (V. pedata), 35.
Birthwort Family, 89.
Bitter Dock, 92.
Bitternut Hickory, 96.
Bittersweet, 85.
Black Alder, 74.
Black Ash, 89.
Blackberry, 49, 50.
Black Bindweed, 92.
Black Birch, 98.
Black Cherry, 47.
Black Currant, 51.
Black Grass, 113.
Black Horehound, 83.
Black Medick, 45.
Black Mustard, 34.
Black Oak, 97.
Black Oat Grass, 124.
Black Raspberry (Thimbleberry), 49.
Black Scrub Oak, 97.
Black Snakeroot, 31, 56.
Black Spruce, 12, 101.
Black Thorn, 50.
Black Walnut, 96.
Bladder Campion, 38.
Bladderwort, 76.
Bladderwort Family, 76.
Bladder Ketmia, 40.
Blazing star, 60.
Blitum, 90.
Blue Beech, 98.
Blueberry, 71.
Bluebottle, 67.
Blue Cohosh, 31.
Blue-curls, 81.
Blue-eyed Grass, 110.
Blue Flag, 110.
Blue Grass, 126.
Blue-joint Grass, 124.
Blue Lettuce, 69.
Blue Vervain, 80.
Blue Violet, 35.

Blue-weed, 83.
Bloodroot, 32.
Bœhmeria, 95.
Borage Family, 83.
Borraginaceæ, 83.
Bostrychia, 158.
Botrychium, 132.
Bottle Brush Grass, 128.
Bouncing Bet, 37.
Bouteloua, 125.
Brake, 133.
Brachyelytrum, 124.
Brassica, 34.
Brasenia, 31.
Bristly Crowfoot, 30.
Bristly Sarsaparilla, 57.
Briza, 127.
Brizopyrum, 126.
Bromus, 127.
Broom-rape Family, 77.
Brown Bent Grass, 123.
Brunella, 82.
Bryaceæ, 138.
Bryeæ, 140.
Bryopsis, 165.
Bryum, 140, 171.
Buckbean, 87.
Buckthorn, 43.
Buckthorn Family, 43.
Buckwheat, 92.
Buckwheat Family, 91, 92.
Buellia, 154.
Bugle-weed, 81.
Bunch Berry, 57.
Burdock, 68.
Bur-reed, 104.
Bush Honeysuckle, 95.
Butter-and-eggs, 78.
Buttercups, 30.
Butterfly-weed, 88.
Butternut, 96.
Button Bush, 60.
Buttonwood, 95.
Buxbaumicæ, 140.

Cabbage, 34.
Cactaceæ, 55.
Cactus Family, 55.
Cakile, 35.
Calla, 103.
Calamagrostis, 124.
Calamagrostis Pickeringii, 12, 124.
Calicium, 155.
Calliblepharis, 159.
Callithamnion, 161.
Callitrichaceæ, 93.
Callitriche, 93.
Calluna, 72.
Calopogon, 109.
Calothrix, 167.
Caltha, 30.
Calystegia, 85.
Camelina, 34.
Campanula, 70.
Campanulaceæ, 70.
Campanula Family, 70.
Canada Thistle, 68.
Canadian Burnet, 48.
Canary Grass, 129.
Cannabis, 95.
Canoe Birch, 98.
Caprifoliaceæ, 58.
Capsella, 34.
Caraway, 56.
Cardamine, 33.
Cardinal Flower, 70.
Carex. 118, 169.
Carpet-weed, 39.
Carrion Flower, 110.
Carrot, 56.
Carum, 56.
Carya, 96.
Caryophyllaceæ, 37.
Cashew Family, 42.
Cassandra, 72.
Cassia, 47.
Castagnea, 164.
Castanea, 97.
Castilleia, 80.

Catalpa, 77.
Catchfly, 38.
Catnip, 82.
Cat-tail, 103.
Cat-tail Family, 103.
Caulophyllum, 31.
Ceanothus, 43.
Celandine, 32.
Celastraceæ, 43.
Celastrus, 43.
Celtis, 94.
Centaurus, 67, 68.
Cephalanthus, 60.
Ceramieæ, 161.
Ceramium, 161.
Cerastium, 38.
Ceratodon, 138.
Ceratophyllaceæ, 93.
Ceratophyllum, 93.
Cetraria, 149.
Chætomorpha, 167.
Chamæcyparis, 101.
Chantransia, 162.
Chara, 147.
Characeæ, 146.
Chareæ, 147.
Charlock, 35.
Cheat, 127.
Checkerberry, 72.
Chelidonium, 32.
Chelone, 78.
Chenilla, 158.
Chenopodiaceæ, 89, 90.
Chenopodina (Suæda), 90.
Chenopodium, 89.
Cherry, 48.
Chervil, 19, 57.
Chess, 127.
Chestnut, 97.
Chestnut Oak, 97.
Chickweed, 38.
Chiloscyphus, 145.
Chimaphila, 74.
Chinese Sumach, 42.

Chinquapin Oak, 12, 97.
Chiogenes, 72.
Chlorosporæ, 165.
Choke-berry, 50.
Choke Cherry, 47.
Chondrus, 160.
Chorda, 164.
Chordaria, 164.
Christmas Fern, 135.
Chrysosplenium, 52.
Chylocladia (Lomentaria), 160.
Cichorium, 68.
Cichory, 68.
Cicuta, 56.
Cimicifuga, 31.
Cinna, 124.
Cinque-foil, 48, 49.
Circæa, 53.
Cirsium, 68.
Cissus Sieboldii, 42.
Cistaceæ, 36.
Cladium, 118.
Cladonia, 153.
Cladophora, 166.
Cladostephus, 164.
Clammy Locust, 45.
Clathrocystis, 168.
Claytonia, 40, 169.
Clearweed, 95.
Clematis, 29.
Clethra, 72.
Climacieæ, 142.
Climacium, 142, 172.
Climbing Bittersweet, 43.
Climbing Fern, 12, 136.
Climbing Fumitory, 32.
Climbing Hemp-weed, 61.
Clintonia, 111.
Closed Gentian, 87.
Clover, 45.
Club Moss Family, 131.
Club Rush, 117.
Clustered Bellflower, 70.
Coast Blite, 90.

Coast Knot-grass, 92.
Cockle-bur, 65.
Collema, 152.
Collinsonia, 82.
Coltsfoot, 61.
Columbine, 30.
Comandra, 93.
Comfrey, 84.
Common Lady's Slipper, 110.
Common Nightshade, 85.
Common Yellow-rattle, 80.
Compositæ, 60.
Composite Family, 60.
Comptonia, 98.
Cone-flower, 65.
Coniferæ, 101.
Conium, 57.
Conocephalus, 144.
Conotrema, 153.
Convallaria, 111.
Convolvulaceæ, 84.
Convolvulus, 84.
Convolvulus Family, 84.
Corallineæ, 159.
Coptis, 30.
Corallina, 159.
Corallineæ, 159.
Corallorhiza, 109.
Coreopsis, 65.
Corn, 130.
Cornaceæ, 57.
Corn Cockle, 38.
Corn Speedwell, 79.
Corn Spurry, 39.
Cornus, 57.
Corydalis, 33.
Corylus, 98.
Cotton Grass, 118.
Cotton Thistle, 68.
Couch Grass, 127.
Cow-berry, 71.
Cow Lily, 32.
Cow Parsnip, 56.
Cow-wheat, 80.

Crab Grass, 129.
Cranberry, 71.
Cranberry Tree, 59.
Crassulaceæ, 52.
Cratægus, 50.
Creeping Crowfoot, 30.
Creeping Snowberry, 72.
Creeping Spearwort, 30.
Creeping Thyme, 81.
Crowfoot Family, 29.
Cruciferæ, 33.
Cryptotæmia, 57.
Cucurbitaceæ, 55.
Cudweed, 67.
Cup Plant, 64.
Cupressus thyoides, 12, 101.
Cupressus (Chamæcyparis), 101.
Cupuliferæ, 96.
Curled Dock, 92.
Curled Mallow, 169.
Currant, 51.
Cursed Crowfoot, 30.
Cuscuta, 85.
Cutler, Manasseh, 15, 19, 20, 169.
Cyanophyceæ, 167.
Cylindrothecieæ, 142.
Cylindrothecium, 142.
Cylindrothecium (Leskea), 172.
Cynoglossum, 84.
Cynosurus, 129.
Cyperaceæ, 115.
Cyperus, 115.
Cypripedium, 109, 169.
Cystoclonium, 160.
Cystopteris, 135.

Dactylis, 125.
Daisy Fleabane, 63.
Dalibarda (Rubus), 49.
Dandelion, 69.
Dangleberry, 71.
Danthonia, 128.
Daphne, 93.
Darnel, 127.

Dasya, 158.
Datura, 86.
Daucus, 56.
Delesseria, 159.
Deptford Pink, 37.
Desmarestia, 165.
Desmodium, 45.
Devil's apron, 163.
Dewberry, 49.
Dianthus, 37.
Diatoms, 168.
Dicentra, 33.
Dichelyma, 141.
Dicksonia, 135.
Dicraneæ, 138.
Dicranum, 138, 170, 171.
Dictyosiphon, 165.
Didyodon, 170.
Diervilla, 59.
Diphyscium, 140, 170.
Diplopappus (Aster), 62.
Dipsaceæ, 60.
Dipsacus, 60, 169.
Distichium (Didymodon), 170.
Ditch Stone-crop, 52.
Dockmakie, 59.
Dodder, 85.
Dogbane Family, 87.
Dog Grass (Couch), 127.
Dog Violet, 35.
Dog's tail Grass, 125.
Dog's tooth Violet, 112.
Dogwood Family, 57.
Dough-balls, 158.
Draba, 34.
Draba Caroliniana, 12, 34.
Drosera, 36.
Droseraceæ, 36.
Drummondia (Orthotrichum), 170.
Duckweed Family, 103.
Dulichium, 116.
Dumontieæ, 161.
Dulse, 160.
Dutchman's Breeches, 33.

Dwarf Blueberry, 71.
Dwarf Cherry, 47.
Dwarf Cornel, 57.
Dwarf Dandelion, 68.
Dwarf Ginseng, 57.
Dwarf Gray Willow, 99.
Dwarf Raspberry, 49.
Dwarf Sumach, 42.

Early Blue Violet, 35.
Early Crowfoot, 30.
Early Meadow Rue, 29.
Early Saxifrage, 51.
Eatonia, 125.
Echinocystis, 55.
Echinodorus, 106.
Echinospermum, 84.
Echium, 83.
Ectocarpus, 164.
Eel-grass, 104.
Elachista, 164.
Elatinaceæ, 37.
Elatine, 37.
Elder, 59.
Elecampane, 64.
Eleusine, 125.
Eleusine Indica, 16, 125.
Elm, 94.
Eleocharis, 116.
Elodes, 37.
Elymus, 128.
Enchanter's Nightshade, 53.
Endocarpon, 155.
Endogens, 103.
English Hawthorn, 50.
English Oak, 97.
English Plantain, 75.
English Violet, 35.
Euteromorpha, 166.
Ephebe, 151.
Epigæa, 72.
Epilobium, 53.
Epiphegus, 77.
Equisetaceæ, 133.

Equisetum, 133.
Eragrostis, 126.
Erechthites, 67.
Ericaceæ, 71.
Erigeron, 62.
Eriocaulon, 114.
Eriocaulonaceæ, 114.
Eriophorum, 117.
Erodium, 41.
Erythronium, 113.
Erythrotrichia, 162.
Euphorbia, 94.
Euphorbiaceæ, 94.
Eupatorium, 60.
European Ivy, 57.
European Linden, 40.
European Mountain Ash, 56.
European Walnut, 96.
European Water Cress, 33.
Euthora, 160.
Evening Lychnis, 38.
Evening Primrose, 54.
Evening Primrose Family, 53.
Evergreen (Lycopodium), 131.
Everlasting, 67.
Evernia, 149.
Exogens, 29.

Fagopyrum, 92.
Fagus, 97.
Fall Dandelion, 69.
False Buckwheat, 92.
False Flax, 34.
False Mitre-wort, 52.
False Nettle, 95.
False Pimpernel, 79.
False Red-top, 126.
False Spikenard, 111.
Featherfoil, 76.
Feather Grass, 124.
Fegatella (Conocephalus), 144.
Ferns, 133.
Festuca, 127.
Feverfew, 56.

Ficoideæ, 39.
Field Chickweed, 38.
Field Pennycress, 34.
Field Sorrel, 92.
Figwort, 78.
Figwort Family, 78.
Filices, 133.
Fimbriaria, 144.
Fimbristylis, 118.
Finger Grass, 129.
Fireweed, 67.
Fissidens, 138.
Fissidenteæ, 138.
Five-finger, 48.
Flax, 41.
Flax Dodder, 85.
Flax Family, 41.
Fleabane, 63.
Floating Foxtail, 123.
Floating Heart, 87.
Florideæ, 158, 162.
Flowering Dogwood, 58.
Flowering Fern, 136.
Flowering Raspberry, 19, 49.
Fly Honeysuckle, 58.
Fontinaleæ, 141.
Fontinalis, 141.
Fool's Parsley, 56.
Forget-me-not, 84.
Forked Chickweed, 39.
Fowl Meadow Grass, 125.
Fragaria, 49.
Fraxinus, 89.
Fringed Gentian, 86.
Fringed Polygala, 44.
Frog's-bit-Family, 107.
Frost Grape, 42.
Frost-weed, 36.
Frullania, 145.
Fucaceæ, 162.
Fucus, 162.
Fumaria, 33.
Fumariaceæ, 32.
Fumitory Family, 32.

Funaria, 141, 171.
Funarieæ, 141.
Fungi, 156.

Galeopsis, 82.
Galium, 59.
Garden Asparagus, 112.
Garden Caraway, 56.
Garden Phlox, 84.
Garden Radish, 35.
Garden Sunflower, 65.
Garden Thyme, 81.
Garden Yellow Honeysuckle, 58.
Garget, 89.
Gaultheria, 72.
Gaylussacia, 71, 169.
Genista, 12, 44.
Gentiana, 86.
Gentianaceæ, 86.
Gentian Family, 86.
Geocalyx, 145.
Geraniaceæ, 41.
Geranium, 41.
Geranium Family, 41.
Gerardia, 79.
Geum, 48.
Gigartina, 160.
Gigartineæ, 160.
Ginseng Family, 57.
Glaux, 76.
Gleditschia, 47.
Gloiosiphonia, 161.
Glyceria, 125.
Gnaphalium, 67.
Golden Moss, 52.
Golden Ragwort, 67.
Golden Saxifrage, 52.
Goldthread, 30.
Goniotrichum, 162.
Goodyera, 108.
Gooseberry, 51.
Goosefoot, 89.
Goosefoot Family, 89.
Gourd Family, 55.

Gracilaria, 159.
Gramineæ, 122.
Grape, 42.
Grape Hyacinth, 112.
Graphis, 155.
Grass Family, 122.
Grass of Parnassus, 51.
Gratiola, 79.
Gray Birch, 98.
Great Angelica, 56.
Great Bulrush, 117.
Great Lobelia, 70.
Great Purple Orchis, 108.
Great Solomon's Seal, 112.
Great Water Dock, 92.
Greenbrier, 110.
Griffithsia, 161.
Grimmia (Schistidium), 139, 170.
Grimmieæ, 139.
Ground Cherry, 85.
Ground Hemlock, 102.
Ground Ivy, 82.
Ground Nut, 46.
Ground Pine, 131.
Groundsel, 67.
Gyalecta, 153.
Gymnogongrus, 160.
Gymnosperms, 101.
Gymnostichum, 128.
Gymnostomum, 170.

Habenaria, 107.
Hackberry, 94.
Hair Grass, 123, 128.
Hairy Sweet Cicely, 57.
Halberd-leaved Tear-Thumb, 92.
Halorageæ, 53.
Halosaccion, 161.
Hamamelaceæ, 52.
Hamamelis, 52.
Hardhack, 48.
Harebell, 70.
Harpanthus, 145.
Hawthorn, 50.

Hay-scented Fern, 135.
Hazelnut, 98.
Heart's-Ease, 36.
Heather, 72.
Heath Family, 71.
Hedeoma, 81.
Hedera, 57.
Hedge Bind-weed, 85.
Hedge Mustard, 33.
Hedge Nettle, 83.
Hedwigia, 139.
Hedwigia (Anictangium), 170.
Hedwigieæ, 139.
Helenium, 66.
Helianthemum, 36.
Helianthus, 65.
Hellebore, 111.
Helminthocladieæ, 160.
Hemerocallis, 113.
Hemlock Spruce, 101.
Hemp, 95.
Hemp Nettle, 82.
Henbane, 86.
Hepatica, 29.
Hepaticæ, 144.
Hepatica triloba (Anemone), 29.
Heracleum, 56.
Herb Robert, 41.
Heterothecium, 154.
Hibiscus, 40.
Hickory, 96.
Hieracium, 69.
Hierochloa, 129.
Higginson, Francis, 19.
High Blackberry, 50.
High-bush Blueberry, 71.
High-water Shrub, 64.
Hildenbrandtia, 159.
Hoary Alder, 99.
Hoary Willow, 99.
Hobble-bush, 59.
Hog Peanut, 47.
Holcus, 128.
Holly Family, 74.

Honey Locust, 47.
Honeysuckle Family, 58.
Honeywort, 57.
Hooked Crowfoot, 30.
Hop, 95.
Hop Hornbeam, 98.
Hordeum, 128.
Horehound, 82.
Hormotrichum, 167.
Hornbeam, 98.
Horned Pond-weed, 104.
Hornwort Family, 93.
Horse Chestnut, 43.
Horseradish, 33.
Horsetail, 133.
Horsetail Family, 133.
Horse-weed, 62.
Hottonia, 76.
Hound's-tongue, 84.
Houseleek, 52.
Houstonia, 60.
Huckleberry, 71.
Hudsonia, 36.
Humulus, 95.
Hungarian Grass, 129.
Hydrocharidaceæ, 107.
Hydrocotyle, 55.
Hyoscyamus, 86.
Hyoscyamus niger, 15, 86.
Hypericaceæ, 37.
Hypericum, 37.
Hypneæ, 142.
Hypnum, 142, 172.
Hypoxys, 110.
Hyssopus, 81.

Ilex, 74.
Ilicineæ, 74.
Ilysanthes, 79.
Impatiens, 41.
Indian Corn, 130.
Indian Cucumber-root, 111.
Indian Grass, 130.
Indian Hemp, 87.

Indian Pipe, 74.
Indian Poke, 111.
Indian Rice, 122.
Indian Tobacco, 70.
Indian Turnip, 103.
Inkberry, 75.
Interrupted Flowering Fern, 136.
Inula, 64.
Ipomœa, 84.
Iridaceæ, 110.
Iris, 110.
Iris Family, 110.
Irish Moss, 160.
Isoetes, 131.
Iva, 64, 169.
Ivy, 42.

Jack-in-the-pulpit, 103.
Jerusalem Artichoke, 65.
Jerusalem Oak, 89.
Jewel-weed (Impatiens), 41.
Jointed Charlock, 35.
Joint-weed, 91.
Josselyn, John, 14, 20.
Juglandaceæ, 96.
Juglans, 96.
Juncaceæ, 113.
Juncus, 113.
June-berry, 51.
June Grass, 126.
Jungermannia, 145.
Jungermanniaceæ, 144.
Juniper, 102.
Juniperus, 102.

Kalmia, 73.
Kelp, 163.
Kentucky Blue Grass, 126.
Knapweed, 68.
Knawel, 39.
Knot-grass, 92.
Krigia, 68.

Labiatæ, 80.

Labrador Tea, 73.
Lactuca, 69.
Lady Fern, 134.
Lady's Slipper, 109.
Lady's Thumb, 91.
Ladies' Tresses, 108.
Lambkill, 73.
Laminaria, 163.
Lamium, 83.
Lampsana, 68.
Lance-leaved Violet, 35.
Laportea, 95.
Lappa, 68.
Larch, 101.
Large Cranberry, 71.
Larger Bur Marigold, 66.
Larger Yellow Lady's Slipper, 109.
Larix, 101.
Lathyrus, 46.
Lauraceæ, 93.
Laurel, 73.
Laurel Family, 93.
Laver, 162.
Leadwort Family, 75.
Leathesia, 164.
Lecanora, 152.
Lechea, 36.
Lecidea, 154.
Ledum, 73.
Leersia, 122.
Leguminosæ, 44.
Lemna, 103.
Lemnaceæ, 103.
Lentibulaceæ, 76.
Leontodon, 69.
Leonurus, 83.
Lepidium, 34.
Lepigonum, 39.
Leptogium, 152.
Leskea, 141, 172.
Lespedeza, 46.
Leucanthemum, 66.
Leucobryeæ, 138.
Leucobryum, 138.

Leucobryum (Dicranum), 170.
Leucodon, 172.
Leucothœ, 72.
Levisticum, 56.
Liatris, 60.
Lichens, 149.
Lichina, 152.
Ligusticum, 56.
Ligustrum, 88.
Lilac, 88.
Liliaceæ, 110.
Lilium, 112.
Lily Family, 110.
Lily-of-the-Valley, 111.
Limanthemum, 87.
Linaceæ, 41.
Linaria, 78.
Linden, 40.
Linden Family, 40.
Lindera, 93.
Linnæa, 58.
Linum, 41, 169.
Liparis, 109.
Liriodendron, 31.
Listera, 108.
Lithothamnion, 159.
Live-for-ever, 52.
Liverwort Family, 144.
Lobelia, 70.
Lobeliaceæ, 70.
Lobelia Family, 70.
Lobster-claws, 158.
Locust, 45.
Lolium, 127.
Lombardy Poplar, 100.
Lomentaria, 160.
Long-pointed Anemone, 29.
Long-fruited Strawberry, 49.
Long-leaved Stitchwort, 38.
Lonicera, 58.
Loosestrife Family, 54.
Lophocolea, 145.
Lopseed, 80.
Lousewort, 80.

Lovage, 56.
Love-lies-bleeding, 15, 91.
Low Birch, 98.
Low Blackberry, 49.
Low Blueberry, 71.
Low Spear Grass, 126.
Low Yellow Clover, 45.
Lucerne, 45.
Ludwigia, 54.
Lunularia, 144.
Lupine, 44.
Lupinus, 44.
Luzula, 113.
Lychnis, 38.
Lycium, 86.
Lycopodiaceæ, 131.
Lycopodium, 131.
Lycopsis, 83.
Lycopus, 81.
Lygodium, 136.
Lygodium palmatum, 12, 136.
Lyngbya, 167.
Lysimachia, 75.
Lysimachia (Steironema), 76.
Lythraceæ, 54.
Lythrum, 55.

Madder Family, 59.
Mad-dog Skull-cap, 82.
Madotheca, 145.
Magnolia, 31.
Magnoliaceæ, 31.
Magnolia Family, 31.
Magnolia Glauca, 11, 31.
Malanthemum, 112.
Maidenhair, 133.
Mallow, 40.
Mallow Family, 40.
Malva, 40, 169.
Malvaceæ, 40.
Mandrake, 31.
Maple-leaved Goosefoot, 89.
Marchantia, 144.
Marchantiaceæ, 144.

Marigold, 66.
Marine Algæ, 157.
Marsh Cress, 33.
Marsh Five-finger, 49.
Marsh Grass, 125.
Marsh Mallow, 40.
Marsh Marigold, 30.
Marsh Rosemary, 75.
Marsh Speedwell, 79.
Marsh St. John's-wort, 37.
Marsh Vetchling, 46.
Marsilia, 132.
Martynia, 77.
Marrubium, 82.
Maruta, 66.
Mastigobryum, 145.
Matricaria (Pyrethrum), 66.
Matrimony Vine, 86.
May Apple, 31.
May-flower, 72.
May-weed, 66.
Meadow Beauty, 54.
Meadow Fescue, 127.
Meadow Fox-tail, 123.
Meadow Grass, 126.
Meadow Parsnip, 56.
Meadow Sweet, 48.
Medeola, 111.
Medicargo, 45.
Melampyrum, 80.
Melanosporæ, 162.
Melastomaceæ, 54.
Melastoma Family, 54.
Melilotus, 45.
Melobesia, 159.
Mentha, 81.
Menyanthes, 87.
Mermaid-weed, 53.
Mesogloia (Castagnea), 164.
Mesogloia vermicularis, 168.
Mezereum Family, 93.
Microstylis, 109.
Mikania, 61.
Mild Water Pepper, 91.

Milkweed Family, 87.
Milkwort Family, 44.
Millet, 129.
Mimulus, 78.
Mint Family, 80.
Missouri Currant, 51.
Mitchella, 60.
Mnium, 140, 169.
Mockernut, 96.
Mock Orange, 51.
Mollugo, 39.
Monarda, 82.
Monardadidyma, 82.
Moneses, 74.
Moneywort, 76.
Monkey Flower, 78.
Monostroma Pulchrum, 168.
Monostroma (Ulva), 166.
Monotropa, 74.
Morning Glory, 84.
Morus, 95.
Mosses, 137.
Mossy Stone-crop, 52.
Motherwort, 83.
Moth Mullein, 78.
Mountain Ash, 50.
Mountain Holly, 75.
Mountain Laurel, 73.
Mountain-mint, 81.
Mouse-ear Chickweed, 38.
Mugwort, 67.
Muhlenbergia, 124.
Mulgidium, 69.
Mullein, 78.
Muscari, 112.
Musci, 137, 170.
Muscineæ, 137.
Musk Mallow, 40.
Mustard Family, 35.
Mycoporum, 155.
Myosotis, 84.
Myrica, 98.
Myricaceæ, 98.
Myrionema, 164.

Myriophyllum, 53.

Nabalus, 69.
Naiadaceæ, 104.
Naias, 104.
Narrow-leaved Chain Fern, 133.
Nasturtium, 33.
Neckera, 142, 172.
Neckereæ, 142.
Negundo, 44.
Nemalion, 160.
Nemopanthes, 75.
Nepeta, 82.
Nephroma, 151.
Nesæa, 55.
Nettle, 95.
Nettle Family, 94.
Nettle Tree, 94.
New Jersey Tea, 43.
New York Fern, 134.
Nicandra, 86.
Nichols, Andrew, 22.
Nicotiana, 86.
Nicotiana rustica, 15, 86.
Nigger-hair, 158.
Night-flowering Catchfly, 38.
Nightshade Family, 85.
Nipplewort, 68.
Nitella, 146.
Nitelleæ, 146.
Nodding Trillium, 111.
Nonesuch, 45.
Northern Fox Grape, 42.
Norway Pine, 101.
Nuphar, 32.
Nymphæa, 32.
Nymphæaceæ, 31.
Nyssa, 58.

Oakes, William, 22.
Oakesia, 111.
Oak Family, 96-98.
Oak-leaved Goosefoot, 89.
Oat, 128.

Œnothera, 54.
Old Witch Grass, 130.
Oleaceæ, 88.
Olive Family, 88.
Onagraceæ, 53.
Onoclea, 135.
Onopordon, 68.
Opegrapha, 154.
Ophioglossaceæ, 132.
Ophioglossum, 132.
Opuntia, 55.
Orange Grass, 37.
Orchard Grass, 125.
Orchidaceæ, 107.
Orchis, 107.
Orchis Family, 107.
Origanum, 81.
Ornithogalum, 112.
Orobanchaceæ, 77.
Orpine Family, 52.
Orthotricheæ, 139.
Orthotrichum, 139, 171.
Oryzopsis, 124.
Osgood, George, 15, 21.
Osier Willow, 99.
Osmorrhiza, 19, 57.
Osmunda, 136.
Ostrich Fern, 135.
Ostrya, 98.
Oswego Tea, 82.
Oxalis, 41.

Painted Cups, 80.
Painted Trillium, 111.
Pale Golden-rod, 63.
Pale Laurel, 73.
Panicum, 129.
Pannaria, 151.
Papaveraceæ, 32.
Paper Birch, 98.
Parmelia, 150.
Parnassia, 51.
Paronychieæ, 39.
Parsley Family, 55.

Parsnip, 56.
Partridge Berry (Mitchella), 60.
Paspalum, 129.
Pastinaca, 56.
Pasture Thistle, 68.
Peach, 48.
Pearlwort, 39.
Pearly Everlasting, 67.
Pedate Violet, 35.
Pedicularis, 80.
Pellia, 144.
Peltandra, 103.
Peltigera, 151.
Penthorum, 52.
Pepper-grass, 34.
Peppermint, 81.
Pertusaria, 153.
Petrocelis, 160.
Petunia, 86.
Phæosporeæ, 163.
Phalaris, 129.
Phegopteris, 134.
Pheum, 123.
Philadelphus, 51.
Phlox, 84.
Phlox Family, 84.
Phragmites, 127.
Phraseolus, 47.
Phryma, 80.
Phyllitis, 165.
Phyllophora, 160.
Physalis, 85.
Physcia, 150.
Physcomitrium, 141.   [170.
Physcomitrium (Gymnostomum),
Physostegia, 82.
Phytolacca, 89.
Phytolaccaceæ, 89.
Picea, 101.
Pickerel-weed, 114.
Pickerel-weed Family, 114.
Pickering, Charles, 23.
Pignut Hickory, 96.
Pigweed, 89.

Pilea, 95.
Pimpernel, 76.
Pine Family, 101.
Pine-sap, 74.
Pine-weed, 37.
Pink Family, 37.
Pinus, 101.
Pinus resinosa, 12, 101.
Pipewort, 114.
Pipewort Family, 114.
Pipsissewa, 74.
Pirus, 50.
Pitcher Plant, 32.
Pitcher Plant Family, 32.
Pitch Pine, 101.
Placodium, 152.
Plagiothecium, 144.
Plane Tree Family, 95.
Plantaginaceæ, 75.
Plantago, 75.
Plantago Maritima, 75.
Plantaio, 75.
Plantain Family, 75.
Plantanaceæ, 95.
Platanus, 95.
Pleurisy-root, 88.
Pluchea, 64.
Plum, 48.
Plumbaginaceæ, 75.
Poa, 126.
Podophyllum, 31.
Pogonatum (Polytrichium), 171.
Pogonia, 108.
Poison Hemlock, 57.
Poison Ivy, 42.
Poison Sumach, 42.
Poke, 89.
Poke Milk-weed, 88.
Poke-weed Family, 89.
Polemoniaceæ, 84.
Polygala, 44.
Polygalaceæ, 44.
Polygonaceæ, 91.
Polygonatum, 112.

Polygonum, 91.
Polygonum Careyi, 12, 91.
Polyides, 160.
Polypodium, 133.
Polypody, 133.
Polypogon, 123.
Polysiphonia, 158.
Polytricheæ, 140.
Polytrichum, 140, 171.
Pond-weed Family, 104.
Pontederia, 114.
Pontederiaceæ, 114.
Poor-man's Weather-glass, 76.
Poplar, 100.
Poppy Family, 32.
Populus, 100.
Porphyra, 162.
Porphyreæ, 162.
Portulaca, 39.
Portulacaceæ, 39.
Potamogeton, 104.
Potentilla, 48.
Potentilla tridentata, 12, 49.
Poterium, 48.
Pottia, 139.
Pottia (Gymnostomum), 170.
Pottieæ, 139.
Poverty Grass, 125.
Prairie Willow, 99.
Preissia, 144.
Prickly Pear, 55.
Prickly Poppy, 32.
Prim, 88.
Primulaceæ, 75.
Primrose Family, 75.
Primrose-leaved Violet, 35.
Prince's Feather, 15, 91.
Prince's Pine, 74.
Privet, 88.
Proserpinaca, 53.
Prunus, 47.
Pterigynandrum, 172.
Pteris, 133.
Ptilidium (Belpharozia,) 145.

Ptilota, 161.
Puke Family, 44.
Punctaria, 165.
Purple Avens, 48.
Purple Azalea, 73.
Purple Orchis, 108.
Purple Thorn-apple, 86.
Purple Trillium, 110.
Purslane, 39.
Purslane Family, 39.
Pussy Willow, 99.
Putty-root, 109.
Pycnanthemum, 81.
Pylaisæa, 142.
Pylaisæa (Pterigynandrum), 172.
Pylaisæeæ, 142.
Pyrenopsis, 152.
Pyrenula, 155.
Pyrethrum, 66.
Pyrola, 73.
Pyxine, 151.

Quaking Grass, 127.
Quercus, 96.
Quercus prinoides, 12, 97.
Quick Grass, 127.
Quillwort, 131.

Rabbit-foot Clover, 45.
Radish, 35.
Radula, 145.
Ralfsia, 164.
Ramalina, 149.
Ranunculaceæ, 29.
Ranunculus, 29.
Ranunculus alismæfolius, 29.
Raphanus, 35.
Raspberry, 49.
Rattlesnake Fern, 132.
Rattlesnake Grass, 125.
Rattlesnake-root, 69.
Rattlesnake-weed, 69.
Red-ash, 89.
Red Baneberry, 31.

Red-berried Elder, 59.
Red Birch, 98.
Red Cedar, 102.
Red Clover, 45.
Red Currant, 51.
Red Hemp Nettle, 82.
Red Maple, 43.
Red Oak, 97.
Red Osier Dogwood, 58.
Red Pine, 12, 101.
Red Plum, 47.
Red Snow, 168.
Red-stemmed Plantain, 75.
Red-top, 123.
Reed Meadow Grass, 126.
Rhabdonia, 160.
Rhamnaceæ, 43.
Rhamnus, 43.
Rhexia, 54.
Rhinanthus, 80.
Rhizocarpeæ, 132.
Rhizoclonium, 168.
Rhododendron, 73.
Rhodomela, 159.
Rhodomeleæ, 158.
Rhodora (Rhododendron), 73.
Rhodymenia, 160.
Rhodymenieæ, 160.
Rhus, 42.
Rhynchospora, 118.
Ribbon Grass, 129.
Ribes, 51.
Rib-grass, 75.
Riccia, 144.
Ricciaceæ, 144.
Rice Cut Grass, 122.
Rich-weed, 82, 95.
Rinodina, 153.
River Birch, 98.
Rivularia, 168.
Robinia, 45.
Robin's Plantain, 63.
Rock Maple, 43.
Rock Rose Family, 36.

Rockweed, 162, 163.
Roman Wormwood, 64.
Rosa, 50.
Rosaceæ, 47.
Rose Acacia, 45.
Rose Family, 47.
Rough Bed-straw, 59.
Rough Meadow Grass, 126.
Round-leaved Cornel, 58.
Round-leaved Violet, 35.
Roxbury Waxwork, 43.
Royal Fern, 136.
Rubiaceæ, 59.
Rubus, 49.
Rubus odoratus, 19, 49.
Rudbeckia, 65.
Rudbeckia hirta, 16, 65.
Rue Anemone, 29.
Rue Family, 42.
Rumex, 92.
Running Swamp Blackberry, 50.
Ruppia, 104.
Rush Family, 113.
Russell, John L., 24.
Rutaceæ, 42.
Rye, 128.

Sagedia, 155.
Sagina, 39.
Sagittaria, 106.
St. John's-wort, 37.
St. John's-wort Family, 37.
Salicaceæ, 99.
Salicornia, 90.
Salix, 99.
Salix candida, 12.
Salix myrtilloides, 12, 99, 100.
Salsola, 90.
Salt-marsh Fleabane, 64.
Salt-marsh Grass, 125.
Saltwort, 90.
Sambucus, 59.
Sandal-wood Family, 93.
Sand Grass, 125.

Sand Spurry, 39.
Sand-wort, 38.
Sanguinaria, 32.
Sanicula, 56.
Santalaceæ, 93.
Sapindaceæ, 43.
Saponaria, 37.
Sarracenia, 32.
Sarraceniaceæ, 32.
Sarsaparilla, 57.
Sassafras, 93.
Saxifraga, 51.
Saxifragaceæ, 51.
Saxifrage, 51, 52.
Saxifrage Family, 51.
Scapania, 145.
Scarlet-fruited Thorn, 50.
Scarlet Oak, 97.
Scheuchzeria, 106.
Schistidium, 139.
Schistidium (Grimmia), 170.
Schollera, 114.
Scirpus, 116.
Scleranthus, 39.
Scotch Lovage, 56.
Scotch Pine, 101.
Scouring Rush, 133.
Scrophularia, 78.
Scrophulariaceæ, 78.
Scutellaria, 82.
Scytosiphon, 165.
Sea Blite, 90.
Sea Cabbage, 166.
Sea Club Rush, 117.
Sea-cocklebur, 65.
Sea-colander, 163.
Sea Rocket, 35.
Seashore Plantain, 75.
Seaside Crowfoot, 30.
Seaside Goldenrod, 63.
Sea Lettuce, 166.
Sea Milkwort, 76.
Seaweeds, 157.
Secale, 128.

Sedge Family, 115.
Sedum, 52.
Seed-box, 54.
Selaginella, 131.
Self Heal, 82.
Sempervivum, 52.
Senecio, 67.
Sensitive Fern, 135.
Sensitive Plant, 47.
Sericocarpus, 61.
Setaria, 130.
Shadbush, 51.
Shagbark Hickory, 96.
Sheep-berry, 59.
Sheep's Fescue, 127.
Sheep Sorrel, 92.
Shepherd's Purse, 34.
Shining Willow, 100.
Shin-leaf, 74.
Showy Orchis, 107.
Shrubby Althæa, 40.
Shrubby Cinque-foil, 49.
Sickle-pod, 33.
Sicyos, 55.
Side-saddle Flower, 32.
Silene, 38.
Silkweed, 87.
Silky Cornel, 58.
Silky Willow, 99.
Silphium, 64.
Silver-weed, 49.
Silvery Cinque-foil, 48.
Siphoneæ, 165.
Sisymbrium, 33.
Sisyrinchium, 110.
Sium, 56.
Skull-cap, 82.
Skunk Cabbage, 103.
Sleepy Catchfly, 38.
Slender Blue Flag, 110.
Slippery Elm, 94.
Small Bedstraw, 59.
Small Bitter Cress, 33.
Small Bugloss, 83.

Small Cranberry, 71.
Smaller Solomon's Seal, 112. [109.
Smaller Yellow Lady's Slipper,
Small-flowered Crowfoot, 30.
Small-flowered Sweet Brier, 50.
Small Magnolia, 31.
Small Nettle, 95.
Small Pansy, 36.
Smilaceæ, 110.
Smilacina, 111.
Smilæcina bifolia, 112.
Smilax, 110.
Smilax Family, 110.
Smooth Sumach, 42.
Smooth Sweet Cicely, 57.
Smooth Winterberry, 74.
Snake-head, 78.
Snapdragon, 78.
Sneeze-weed, 66.
Sneezewort, 66.
Snowberry, 58.
Soapberry Family, 43.
Soapwort, 37.
Soft Chess, 127.
Soft Rush, 113.
Solanaceæ, 85.
Solanum, 85.
Solidago, 63.
Solieria (Rhabdonia), 160.
Solomon's Seal, 112.
Sonchus, 69.
Sorghum, 130.
Sorrel, 41.
Sow Thistle, 69.
Sparganium, 104.
Spartina, 125.
Spearmint, 81.
Speckled Alder, 99.
Specularia, 71.
Speedwell, 79.
Spergula, 39.
Spergularia (Lepigonum), 39.
Sphacelaria, 164.
Sphærococcoideæ, 159.

Sphagnaceæ, 137.
Sphagnum, 137, 170.
Spice Bush, 93.
Spiked Loosestrife, 55.
Spikenard, 57.
Spiræa, 48.
Spiranthes, 108.
Splachnum, 170.
Spongiocarpeæ, 160.
Sporobolus, 123.
Spotted Cowbane, 56.
Spotted Wintergreen, 74.
Spreading Dogbane, 87.
Spring Beauty, 40.
Spring Clotbur, 65.
Spring Cress, 33.
Spurge Family, 94.
Spyridia, 161.
Spyridieæ, 161.
Squamarieæ, 160.
Squirrel-tail Grass, 128.
Stachys, 83.
Staff Tree Family, 43.
Staghorn Sumach, 42.
Star Cucumber, 55.
Star-flower, 75.
Star-flower (Houstonia), 60.
Star Grass, 110.
Statice, 75.
Steironema, 76.
Stellaria, 38.
Stereocaulon, 153.
Stipa, 124.
Stitchwort, 38.
Stone-crop, 52.
Stone-root, 82.
Storksbill, 41.
Stramonium, 86.
Strawberry, 49.
Strawberry Blite, 90.
Strawberry Tomato, 85.
Streptopus, 111.
Stricta, 151.
Striped Grass, 129.

Striped Maple, 43.
Struthiopteris (Onoclea), 135.
Suæda, 90.
Sugar Maple, 43.
Sumach, 42.
Summary, 173.
Summer Grape, 42.
Sundew Family, 36.
Sundrops, 54.
Sunflower, 65.
Swamp Beggar-ticks, 66.
Swamp Blueberry, 71.
Swamp Hickory, 96.
Swamp Honeysuckle, 73.
Swamp Loosestrife, 55.
Swamp Maple, 43.
Swamp Milk-weed, 88.
Swamp Rose, 50.
Swamp Rose Mallow, 40.
Swamp Saxifrage, 51.
Swamp Thistle, 68.
Swamp Violet, 35.
Swamp White Oak, 97.
Sweet Alyssum, 34.
Sweet Birch, 98.
Sweet Brier, 50.
Sweet Cicely, 19, 57.
Sweet Fern, 98.
Sweet Flag, 103.
Sweet Gale, 98.
Sweet Gale Family, 98.
Sweet Goldenrod, 64.
Sweet-scented Bedstraw, 59.
Sweet Vernal Grass, 129.
Sweet White Violet, 35.
Sweet William, 37.
Sweet William Catchfly, 38.
Symphitum, 83.
Symphoricarpus, 58.
Symplocarpus, 103.
Syringa, 88.

Tall Buttercups, 30.
Tall Meadow Rue, 29.

Tamarack, 101.
Tanacetum, 66.
Tansy, 66.
Taraxacum, 69.
Tare, 46.
Taxodium, 102.
Taxus, 102.
Teasel, 60, 169.
Teasel Family, 60.
Tecoma, 77.
Tephrosia, 45.
Tetraphideæ, 139.
Tetraphis, 139, 170.
Teucrium, 80.
Thalaspi, 34.
Thalictrum, 29.
Thallophytes, 149.
Thaspium, 56.
Thelia, 141.
Thelieæ, 141.
Theloschistes, 150.
Thimbleberry, 49.
Thin Grass, 123.
Thistle, 68.
Thoroughwort, 61.
Three-seeded Mercury, 94.
Three-thorned Acacia, 47.
Thyme, 81.
Thymeleaceæ, 93.
Thyme-leaved Sandwort, 38.
Thymus, 81.
Tiarella, 52.
Tick-seed Sunflower, 65.
Tiger Lily, 112.
Tilia, 40.
Tiliaceæ, 40.
Timothy, 123.
Toad Flax, 78.
Tower Mustard, 33.
Tracy, Cyrus M., 25.
Tree-of-Heaven, 42.
Trichocolea, 145.
Trichostema, 81.
Trichostomeæ, 139.

Trichostomum, 139.
Tricuspis, 125.
Trientalis, 75.
Trifolium, 45.
Triglochin, 106.
Trillium, 110.
Triosteum, 59.
Triticum, 127.
Trumpet Flower, 77.
Trumpet Honeysuckle, 58.
Trypethelium, 155.
Tsuga, 101.
Tufted Loosestrife, 75.
Tulip Tree, 31, 169.
Turk's Cap Lily, 112.
Tupelo, 58.
Turtle-head, 78.
Tussilago, 61.
Tway-blade, 109.
Twisted-stalk, 111.
Typha, 103.
Typhaceæ, 103.

Ulmus, 94.
Ulva, 166.
Umbelliferæ, 55.
Umbilicaria, 151.
Upland Boneset, 61.
Upright Chess, 127.
Urceolaria, 153.
Urtica, 95.
Urticaceæ, 94.
Usnea, 150.
Utricularia, 76.
Uvularia, 111.

Vaccinium, 71.
Vaccinium, Vitis-Idæa, 12, 71.
Vallisneria, 107.
Vanilla Grass, 129.
Vascular Cryptogams, 131.
Vaucheria, 166.
Velvet Grass, 128.
Velvet Leaf, 40.

Wild Gooseberry, 51.
Wild Indigo, 47.
Wild Lettuce, 69.
Wild Lily-of-the-valley, 112.
Wild Liquorice, 60.
Wild Lupine, 44.
Wild Marjoram, 81.
Wild-mint, 81.
Wild Orange-red Lily, 112.
Wild Pepper-grass, 34.
Wild Pink, 38.
Wild Plum, 47.
Wild Red Cherry, 47.
Wild Red Raspberry, 49.
Wild Rose, 50.
Wild Sarsaparilla, 57.
Wild Senna, 47.
Wild Teasel, 60, 169.
Wild Tobacco, 86.
Wild Yellow Lily, 112.
Wild Yellow Plum, 47.
Willow Family, 99.
Willow Herb, 53.
Wire Grass, 126.
Witch Grass (Couch), 127.
Witch Hazel, 52.
Witch Hazel Family, 52.
Withe-rod, 59. ·
Wood Anemone, 29.
Wood Nettle, 95.
Wood Sage, 80.
Wood Waxen, 44.
Woodsia, 135.
Wood Sorrel, 41.

Woodwardia, 133.
Wood, William, 19.
Wormseed, 90.
Wormwood, 15, 67.

Xanthium, 65.
Xanthoxylum, 42.
Xylographa, 155.
Xyridaceæ, 114.
Xyris, 114.

Yarrow, 66.
Yellow Adder's tongue, 112.
Yellow-barked Oak, 97.
Yellow Birch, 98.
Yellow Clover, 45.
Yellow-eyed Grass Family, 114.
Yellow Fringed Orchis, 108.
Yellow Garden Lily, 113.
Yellow Melilot, 45.
Yellow Pond Lily, 32.
Yellow Rocket, 33.
Yellow Thistle, 68.
Yellow Violet, 36.
Yellow Water Crowfoot, 29.
Yellow Wood Sorrel, 41.

Zannichellia, 104.
Zea, 130.
Zigzag Clover, 45.
Zizania, 122.
Zoösporeæ, 166.
Zostera, 104.

www.ingramcontent.com/pod-product-compliance
Lightning Source LLC
Chambersburg PA
CBHW020926230426
43666CB00008B/1592